Student Solution Manual for McKeague's
Introductory Algebra: Concepts and Graphs

Prepared by

Ross Rueger

Department of Mathematics

College of the Sequoias

Visalia, California

Student Solutions Manual for McKeague's
Introductory Algebra: Concepts and Graphs

Ross Rueger

Publisher: XYZ Textbooks

Sales: Amy Jacobs, Richard Jones, Bruce Spears,
    Rachael Hillman

Cover Design: Kyle Schoenberger

Printed in the United States of America

ISBN-13: 978-1-936368-31-0   /   ISBN-10: 1-936368-31-5
XYZ Textbooks

For product information and technology assistance, contact us at
XYZ Textbooks, 1-877-745-3499

For permission to use material from this text or product,
e-mail: **info@mathtv.com**

1339 Marsh Street
San Luis Obispo, CA 93401
USA

For your course and learning solutions, visit **www.xyztextbooks.com**

Contents

Preface

This *Student Solutions Manual* contains complete solutions to all odd-numbered exercises of *Introductory Algebra: Concepts and Graphs* by Charles P. McKeague. Every attempt has been made to format solutions for readability and accuracy, and I apologize for any errors that you may encounter. If you have any comments, suggestions, error corrections, or alternative solutions please feel free to drop me a note or send an email (address below).

Please use this manual with some degree of caution. Be sure that you have attempted a solution, and re-attempted it, before you look it up in this manual. Algebra can only be learned by *doing*, and not by observing! As you use this manual, do not just read the solution but work it along with the manual, using my solution to check your work. If you use this manual in that fashion then it should be helpful to you as you do homework and study for tests.

I wish to express my appreciation to Pat McKeague at MathTV.com and Patrick McKeague at XYZ Textbooks for asking me to be involved with this textbook. This book provides a complete and affordable course in beginning algebra, and you will find the text very easy to read and understand. Good luck!

Ross Rueger
College of the Sequoias
rossrueger@gmail.com

January 2011

Chapter 1
The Basics

1.1 Notation and Symbols

1. The equivalent expression is $x + 5 = 14$.

3. The equivalent expression is $5y < 30$.

5. The equivalent expression is $5y \geq y - 16$.

7. The equivalent expression is $\dfrac{x}{3} = x + 2$.

9. Expanding the expression: $3^2 = 3 \cdot 3 = 9$

11. Expanding the expression: $7^2 = 7 \cdot 7 = 49$

13. Expanding the expression: $2^3 = 2 \cdot 2 \cdot 2 = 8$

15. Expanding the expression: $4^3 = 4 \cdot 4 \cdot 4 = 64$

17. Expanding the expression: $2^4 = 2 \cdot 2 \cdot 2 \cdot 2 = 16$

19. Expanding the expression: $10^2 = 10 \cdot 10 = 100$

21. Expanding the expression: $11^2 = 11 \cdot 11 = 121$

23. Using the order of operations: $2 \cdot 3 + 5 = 6 + 5 = 11$

25. Using the order of operations: $2(3+5) = 2(8) = 16$

27. Using the order of operations: $5 + 2 \cdot 6 = 5 + 12 = 17$

29. Using the order of operations: $(5+2) \cdot 6 = 7 \cdot 6 = 42$

31. Using the order of operations: $5 \cdot 4 + 5 \cdot 2 = 20 + 10 = 30$

33. Using the order of operations: $5(4+2) = 5(6) = 30$

35. Using the order of operations: $8 + 2(5+3) = 8 + 2(8) = 8 + 16 = 24$

37. Using the order of operations: $(8+2)(5+3) = (10)(8) = 80$

39. Using the order of operations: $20 + 2(8-5) + 1 = 20 + 2(3) + 1 = 20 + 6 + 1 = 27$

41. Using the order of operations:
$$5 + 2(3 \cdot 4 - 1) + 8 = 5 + 2(12 - 1) + 8 = 5 + 2(11) + 8 = 5 + 22 + 8 = 35$$

43. Using the order of operations: $8 + 10 \div 2 = 8 + 5 = 13$

45. Using the order of operations: $4 + 8 \div 4 - 2 = 4 + 2 - 2 = 4$

47. Using the order of operations: $3 + 12 \div 3 + 6 \cdot 5 = 3 + 4 + 30 = 37$

49. Using the order of operations: $3 \cdot 8 + 10 \div 2 + 4 \cdot 2 = 24 + 5 + 8 = 37$

51. Using the order of operations: $(5+3)(5-3) = (8)(2) = 16$

53. Using the order of operations: $5^2 - 3^2 = 5 \cdot 5 - 3 \cdot 3 = 25 - 9 = 16$

55. Using the order of operations: $(4+5)^2 = 9^2 = 9 \cdot 9 = 81$

57. Using the order of operations: $4^2 + 5^2 = 4 \cdot 4 + 5 \cdot 5 = 16 + 25 = 41$

59. Using the order of operations: $3 \cdot 10^2 + 4 \cdot 10 + 5 = 300 + 40 + 5 = 345$

61. Using the order of operations: $2 \cdot 10^3 + 3 \cdot 10^2 + 4 \cdot 10 + 5 = 2000 + 300 + 40 + 5 = 2,345$

63. Using the order of operations: $10 - 2(4 \cdot 5 - 16) = 10 - 2(20 - 16) = 10 - 2(4) = 10 - 8 = 2$

65. Using the order of operations:
$$4\left[7+3\left(2\bullet9-8\right)\right]=4\left[7+3\left(18-8\right)\right]=4\left[7+3\left(10\right)\right]=4\left(7+30\right)=4\left(37\right)=148$$

67. Using the order of operations: $5\left(7-3\right)+8\left(6-4\right)=5\left(4\right)+8\left(2\right)=20+16=36$

69. Using the order of operations:
$$3\left(4\bullet5-12\right)+6\left(7\bullet6-40\right)=3\left(20-12\right)+6\left(42-40\right)=3\left(8\right)+6\left(2\right)=24+12=36$$

71. Using the order of operations: $3^4+4^2\div2^3-5^2=81+16\div8-25=81+2-25=58$

73. Using the order of operations: $5^2+3^4\div9^2+6^2=25+81\div81+36=25+1+36=62$

75. Using the order of operations: $20\div2\bullet10=10\bullet10=100$

77. Using the order of operations: $24\div8\bullet3=3\bullet3=9$

79. Using the order of operations: $36\div6\bullet3=6\bullet3=18$

81. Using the order of operations: $48\div12\bullet2=4\bullet2=8$

83. Using the order of operations: $16-8+4=8+4=12$

85. Using the order of operations: $24-14+8=10+8=18$

87. Using the order of operations: $36-6+12=30+12=42$

89. Using the order of operations: $48-12+17=36+17=53$

91. There are $5\bullet2=10$ cookies in the package.

93. The total number of calories is $210\bullet2=420$ calories.

95. There are $7\bullet32=224$ chips in the bag.

97. For a person eating 3,000 calories per day, the recommended amount of fat would be $80+15=95$ grams.

99.   **a.**   The amount of caffeine is: $6\left(100\right)=600$ mg

     b.   The amount of caffeine is: $2\left(45\right)+3\left(47\right)=90+141=231$ mg

101. Completing the table:

Activity	Calories burned in 1 hour
Bicycling	374
Bowling	265
Handball	680
Jogging	680
Skiing	544

103. The pattern is to add 1, so the next number is 5.

105. The pattern is to add 2, so the next number is 10.

107. The next number is $5^2=25$.

109. Since $2+2=4$ and $2+4=6$, the next number is $4+6=10$.

1.2 Real Numbers

1. Labeling the point:

3. Labeling the point:

5. Labeling the point:

7. Labeling the point:

9. Building the fraction: $\dfrac{3}{4}=\dfrac{3}{4}\cdot\dfrac{6}{6}=\dfrac{18}{24}$

11. Building the fraction: $\dfrac{1}{2}=\dfrac{1}{2}\cdot\dfrac{12}{12}=\dfrac{12}{24}$

13. Building the fraction: $\dfrac{5}{8}=\dfrac{5}{8}\cdot\dfrac{3}{3}=\dfrac{15}{24}$

15. Building the fraction: $\dfrac{3}{5}=\dfrac{3}{5}\cdot\dfrac{12}{12}=\dfrac{36}{60}$

17. Building the fraction: $\dfrac{11}{30}=\dfrac{11}{30}\cdot\dfrac{2}{2}=\dfrac{22}{60}$

19. The opposite of 10 is –10, the reciprocal is $\dfrac{1}{10}$, and the absolute value is $|10|=10$.

21. The opposite of $\dfrac{3}{4}$ is $-\dfrac{3}{4}$, the reciprocal is $\dfrac{4}{3}$, and the absolute value is $\left|\dfrac{3}{4}\right|=\dfrac{3}{4}$.

23. The opposite of $\dfrac{11}{2}$ is $-\dfrac{11}{2}$, the reciprocal is $\dfrac{2}{11}$, and the absolute value is $\left|\dfrac{11}{2}\right|=\dfrac{11}{2}$.

25. The opposite of –3 is 3, the reciprocal is $-\dfrac{1}{3}$, and the absolute value is $|-3|=3$.

27. The opposite of $-\dfrac{2}{5}$ is $\dfrac{2}{5}$, the reciprocal is $-\dfrac{5}{2}$, and the absolute value is $\left|-\dfrac{2}{5}\right|=\dfrac{2}{5}$.

29. The opposite of x is $-x$, the reciprocal is $\dfrac{1}{x}$, and the absolute value is $|x|$.

31. The correct symbol is <: $-5<-3$ **33.** The correct symbol is >: $-3>-7$

35. Since $|-4|=4$ and $-|-4|=-4$, the correct symbol is >: $|-4|>-|-4|$

37. Since $-|-7|=-7$, the correct symbol is >: $7>-|-7|$

39. The correct symbol is <: $-\dfrac{3}{4}<-\dfrac{1}{4}$ **41.** The correct symbol is <: $-\dfrac{3}{2}<-\dfrac{3}{4}$

43. Simplifying the expression: $|8-2|=|6|=6$

45. Simplifying the expression: $|5\cdot 2^{3}-2\cdot 3^{2}|=|5\cdot 8-2\cdot 9|=|40-18|=|22|=22$

47. Simplifying the expression: $|7-2|-|4-2|=|5|-|2|=5-2=3$

49. Simplifying the expression:
$$10-|7-2(5-3)|=10-|7-2(2)|=10-|7-4|=10-|3|=10-3=7$$

51. Simplifying the expression:

$$15 - |8 - 2(3 \cdot 4 - 9)| - 10 = 15 - |8 - 2(12 - 9)| - 10$$
$$= 15 - |8 - 2(3)| - 10$$
$$= 15 - |8 - 6| - 10$$
$$= 15 - |2| - 10$$
$$= 15 - 2 - 10$$
$$= 3$$

53. Multiplying the fractions: $\dfrac{2}{3} \cdot \dfrac{4}{5} = \dfrac{8}{15}$

55. Multiplying the fractions: $\dfrac{1}{2}(3) = \dfrac{1}{2} \cdot \dfrac{3}{1} = \dfrac{3}{2}$

57. Multiplying the fractions: $\dfrac{1}{4}(5) = \dfrac{1}{4} \cdot \dfrac{5}{1} = \dfrac{5}{4}$

59. Multiplying the fractions: $\dfrac{4}{3} \cdot \dfrac{3}{4} = \dfrac{12}{12} = 1$

61. Multiplying the fractions: $6\left(\dfrac{1}{6}\right) = \dfrac{6}{1} \cdot \dfrac{1}{6} = \dfrac{6}{6} = 1$

63. Multiplying the fractions: $3 \cdot \dfrac{1}{3} = \dfrac{3}{1} \cdot \dfrac{1}{3} = \dfrac{3}{3} = 1$

65. Expanding the exponent: $\left(\dfrac{3}{4}\right)^2 = \dfrac{3}{4} \cdot \dfrac{3}{4} = \dfrac{9}{16}$

67. Expanding the exponent: $\left(\dfrac{2}{3}\right)^3 = \dfrac{2}{3} \cdot \dfrac{2}{3} \cdot \dfrac{2}{3} = \dfrac{8}{27}$

69. Expanding the exponent: $\left(\dfrac{1}{10}\right)^4 = \dfrac{1}{10} \cdot \dfrac{1}{10} \cdot \dfrac{1}{10} \cdot \dfrac{1}{10} = \dfrac{1}{10,000}$

71. The next number is $\dfrac{1}{9}$.

73. The next number is $\dfrac{1}{5^2} = \dfrac{1}{25}$.

75. The perimeter is $4(1 \text{ in.}) = 4 \text{ in.}$, and the area is $(1 \text{ in.})^2 = 1 \text{ in.}^2$.

77. The perimeter is $2(1.5 \text{ in.}) + 2(0.75 \text{ in.}) = 3.0 \text{ in.} + 1.5 \text{ in.} = 4.5 \text{ in.}$, and the area is $(1.5 \text{ in.})(0.75 \text{ in.}) = 1.125 \text{ in.}^2$.

79. The perimeter is $2.75 \text{ cm} + 4 \text{ cm} + 3.5 \text{ cm} = 10.25 \text{ cm}$, and the area is $\dfrac{1}{2}(4 \text{ cm})(2.5 \text{ cm}) = 5.0 \text{ cm}^2$.

81. A loss of 8 yards corresponds to –8 on a number line. The total yards gained corresponds to –2 yards.

83. The temperature can be represented as –64°. The new (warmer) temperature corresponds to –54°.

85. The wind chill temperature is –15°.

87. His position corresponds to –100 feet. His new (deeper) position corresponds to –105 feet.

89. The area is given by: $8\dfrac{1}{2} \cdot 11 = \dfrac{17}{2} \cdot \dfrac{11}{1} = \dfrac{187}{2} = 93.5 \text{ in.}^2$

    The perimeter is given by: $2\left(8\dfrac{1}{2}\right) + 2(11) = 17 + 22 = 39 \text{ in.}$

91. The calories consumed would be: $2(544) + 299 = 1,387$ calories
93. The calories consumed by the 180 lb person would be $3(653) = 1,959$ calories, while the calories consumed by the 120 lb person would be $3(435) = 1,305$ calories. Thus the 180 lb person consumed $1959 - 1306 = 654$ more calories.
95. **a.**   This is $3 \cdot 31 = 93$ million phones.
     b.   The chart shows $4.5 \cdot 31 = 139.5$ million phones, so this statement is false.
     c.   The chart shows $9.5 \cdot 31 = 294.5$ million phones, so this statement is true.

1.3 Addition of Real Numbers

1.   Adding all positive and negative combinations of 3 and 5:
$$3 + 5 = 8$$
$$3 + (-5) = -2$$
$$-3 + 5 = 2$$
$$(-3) + (-5) = -8$$
3.   Adding all positive and negative combinations of 15 and 20:
$$15 + 20 = 35$$
$$15 + (-20) = -5$$
$$-15 + 20 = 5$$
$$(-15) + (-20) = -35$$
5.   Adding the numbers: $6 + (-3) = 3$
7.   Adding the numbers: $13 + (-20) = -7$
9.   Adding the numbers: $18 + (-32) = -14$
11. Adding the numbers: $-6 + 3 = -3$
13. Adding the numbers: $-30 + 5 = -25$
15. Adding the numbers: $-6 + (-6) = -12$
17. Adding the numbers: $-9 + (-10) = -19$
19. Adding the numbers: $-10 + (-15) = -25$
21. Performing the additions: $5 + (-6) + (-7) = 5 + (-13) = -8$
23. Performing the additions: $-7 + 8 + (-5) = -12 + 8 = -4$
25. Performing the additions: $5 + [6 + (-2)] + (-3) = 5 + 4 + (-3) = 9 + (-3) = 6$
27. Performing the additions: $[6 + (-2)] + [3 + (-1)] = 4 + 2 = 6$
29. Performing the additions: $20 + (-6) + [3 + (-9)] = 20 + (-6) + (-6) = 20 + (-12) = 8$
31. Performing the additions: $-3 + (-2) + [5 + (-4)] = -3 + (-2) + 1 = -5 + 1 = -4$
33. Performing the additions: $(-9 + 2) + [5 + (-8)] + (-4) = -7 + (-3) + (-4) = -14$
35. Performing the additions: $[-6 + (-4)] + [7 + (-5)] + (-9) = -10 + 2 + (-9) = -19 + 2 = -17$
37. Performing the additions: $(-6 + 9) + (-5) + (-4 + 3) + 7 = 3 + (-5) + (-1) + 7 = 10 + (-6) = 4$
39. Using order of operations: $-5 + 2(-3 + 7) = -5 + 2(4) = -5 + 8 = 3$
41. Using order of operations: $9 + 3(-8 + 10) = 9 + 3(2) = 9 + 6 = 15$
43. Using order of operations:
$$-10 + 2(-6 + 8) + (-2) = -10 + 2(2) + (-2) = -10 + 4 + (-2) = -12 + 4 = -8$$
45. Using order of operations: $2(-4 + 7) + 3(-6 + 8) = 2(3) + 3(2) = 6 + 6 = 12$
47. The pattern is to add 5, so the next two terms are $18 + 5 = 23$ and $23 + 5 = 28$.
49. The pattern is to add 5, so the next two terms are $25 + 5 = 30$ and $30 + 5 = 35$.
51. The pattern is to add –5, so the next two terms are $5 + (-5) = 0$ and $0 + (-5) = -5$.

53. The pattern is to add –6, so the next two terms are $-6 + (-6) = -12$ and $-12 + (-6) = -18$.

55. The pattern is to add –4, so the next two terms are $0 + (-4) = -4$ and $-4 + (-4) = -8$.

57. Yes, since each successive odd number is 2 added to the previous one.

59. The expression is: $5 + 9 = 14$

61. The expression is: $[-7 + (-5)] + 4 = -12 + 4 = -8$

63. The expression is: $[-2 + (-3)] + 10 = -5 + 10 = 5$

65. The number is 3, since $-8 + 3 = -5$.   **67.** The number is –3, since $-6 + (-3) = -9$.

69. The expression is $-12° + 4° = -8°$.

71. The expression is $\$10 + (-\$6) + (-\$8) = \$10 + (-\$14) = -\$4$.

73. The new balance is $-\$30 + \$40 = \$10$.

75. The profit in 2007 was: $\$11,500 - \$9,500 = \$2,000$.

77. In the year 2006 the company had more costs than revenue.

1.4 Subtraction of Real Numbers

1. Subtracting the numbers: $5 - 8 = 5 + (-8) = -3$

3. Subtracting the numbers: $3 - 9 = 3 + (-9) = -6$

5. Subtracting the numbers: $5 - 5 = 5 + (-5) = 0$

7. Subtracting the numbers: $-8 - 2 = -8 + (-2) = -10$

9. Subtracting the numbers: $-4 - 12 = -4 + (-12) = -16$

11. Subtracting the numbers: $-6 - 6 = -6 + (-6) = -12$

13. Subtracting the numbers: $-8 - (-1) = -8 + 1 = -7$

15. Subtracting the numbers: $15 - (-20) = 15 + 20 = 35$

17. Subtracting the numbers: $-4 - (-4) = -4 + 4 = 0$

19. Using order of operations: $3 - 2 - 5 = 3 + (-2) + (-5) = 3 + (-7) = -4$

21. Using order of operations: $9 - 2 - 3 = 9 + (-2) + (-3) = 9 + (-5) = 4$

23. Using order of operations: $-6 - 8 - 10 = -6 + (-8) + (-10) = -24$

25. Using order of operations: $-22 + 4 - 10 = -22 + 4 + (-10) = -32 + 4 = -28$

27. Using order of operations: $10 - (-20) - 5 = 10 + 20 + (-5) = 30 + (-5) = 25$

29. Using order of operations: $8 - (2 - 3) - 5 = 8 - (-1) - 5 = 8 + 1 + (-5) = 9 + (-5) = 4$

31. Using order of operations: $7 - (3 - 9) - 6 = 7 - (-6) - 6 = 7 + 6 + (-6) = 13 + (-6) = 7$

33. Using order of operations: $5 - (-8 - 6) - 2 = 5 - (-14) - 2 = 5 + 14 + (-2) = 19 + (-2) = 17$

35. Using order of operations: $-(5 - 7) - (2 - 8) = -(-2) - (-6) = 2 + 6 = 8$

37. Using order of operations: $-(3 - 10) - (6 - 3) = -(-7) - 3 = 7 + (-3) = 4$

39. Using order of operations: $16 - [(4 - 5) - 1] = 16 - (-1 - 1) = 16 - (-2) = 16 + 2 = 18$

41. Using order of operations: $5 - [(2 - 3) - 4] = 5 - (-1 - 4) = 5 - (-5) = 5 + 5 = 10$

43. Using order of operations:

$$21 - \left[-(3-4)-2\right] - 5 = 21 - \left[-(-1)-2\right] - 5$$
$$= 21 - (1-2) - 5$$
$$= 21 - (-1) - 5$$
$$= 21 + 1 + (-5)$$
$$= 22 + (-5)$$
$$= 17$$

45. Using order of operations: $2 \bullet 8 - 3 \bullet 5 = 16 - 15 = 16 + (-15) = 1$

47. Using order of operations: $3 \bullet 5 - 2 \bullet 7 = 15 - 14 = 15 + (-14) = 1$

49. Using order of operations: $5 \bullet 9 - 2 \bullet 3 - 6 \bullet 2 = 45 - 6 - 12 = 45 + (-6) + (-12) = 45 + (-18) = 27$

51. Using order of operations: $3 \bullet 8 - 2 \bullet 4 - 6 \bullet 7 = 24 - 8 - 42 = 24 + (-8) + (-42) = 24 + (-50) = -26$

53. Using order of operations: $2 \bullet 3^2 - 5 \bullet 2^2 = 2 \bullet 9 - 5 \bullet 4 = 18 - 20 = 18 + (-20) = -2$

55. Using order of operations: $4 \bullet 3^3 - 5 \bullet 2^3 = 4 \bullet 27 - 5 \bullet 8 = 108 - 40 = 108 + (-40) = 68$

57. Writing the expression: $-7 - 4 = -7 + (-4) = -11$

59. Writing the expression: $12 - (-8) = 12 + 8 = 20$

61. Writing the expression: $-5 - (-7) = -5 + 7 = 2$

63. Writing the expression: $\left[4 + (-5)\right] - 17 = -1 - 17 = -1 + (-17) = -18$

65. Writing the expression: $8 - 5 = 8 + (-5) = 3$

67. Writing the expression: $-8 - 5 = -8 + (-5) = -13$

69. Writing the expression: $8 - (-5) = 8 + 5 = 13$

71. The number is 10, since $8 - 10 = 8 + (-10) = -2$.

73. The number is –2, since $8 - (-2) = 8 + 2 = 10$.

75. The expression is $\$1,500 - \$730 = \$770$.

77. The expression is $-\$35 + \$15 - \$20 = -\$35 + (-\$20) + \$15 = -\$55 + \$15 = -\$40$.

79. The expression is $73° + 10° - 8° = 83° - 8° = 75°$.

81. The sequence of values is $4500, $3950, $3400, $2850, and $2300. This is an arithmetic sequence, since –$550 is added to each value to obtain the new value.

83. The angles add to $90°$, so $x = 90° - 55° = 35°$.

85. The angles add to $180°$, so $x = 180° - 120° = 60°$.

87.   a.   Completing the table:

Day	Plant Height (inches)
0	0
2	1
4	3
6	6
8	13
10	23

b.   Subtracting: $13 - 1 = 12$. The grass is 12 inches higher after 8 days than after 2 days.

89. a. Yes, the numbers appear to be rounded to the hundreds place.
    b. Subtracting: $21,400 - 13,000 = 8,400$. There were 8,400 more participants in 2004 than in 2000.
    c. If the amount increases by another 8,400, there will be $21,400 + 8,400 = 29,800$ participants in 2008.

1.5 Properties of Real Numbers

1. commutative property (of addition)   3. multiplicative inverse property
5. commutative property (of addition)   7. distributive property
9. commutative and associative properties (of addition)
11. commutative and associative properties (of addition)
13. commutative property (of addition)
15. commutative and associative properties (of multiplication)
17. commutative property (of multiplication)   19. additive inverse property
21. The expression should read $3(x+2) = 3x+6$.
23. The expression should read $9(a+b) = 9a+9b$.
25. The expression should read $3(0) = 0$.   27. The expression should read $3+(-3) = 0$.
29. The expression should read $10(1) = 10$.
31. Simplifying the expression: $4+(2+x) = (4+2)+x = 6+x$
33. Simplifying the expression: $(x+2)+7 = x+(2+7) = x+9$
35. Simplifying the expression: $3(5x) = (3 \bullet 5)x = 15x$
37. Simplifying the expression: $9(6y) = (9 \bullet 6)y = 54y$
39. Simplifying the expression: $\dfrac{1}{2}(3a) = \left(\dfrac{1}{2} \bullet 3\right)a = \dfrac{3}{2}a$

41. Simplifying the expression: $\dfrac{1}{3}(3x) = \left(\dfrac{1}{3} \bullet 3\right)x = 1x = x$

43. Simplifying the expression: $\dfrac{1}{2}(2y) = \left(\dfrac{1}{2} \bullet 2\right)y = 1y = y$

45. Simplifying the expression: $\dfrac{3}{4}\left(\dfrac{4}{3}x\right) = \left(\dfrac{3}{4} \bullet \dfrac{4}{3}\right)x = 1x = x$

47. Simplifying the expression: $\dfrac{6}{5}\left(\dfrac{5}{6}a\right) = \left(\dfrac{6}{5} \bullet \dfrac{5}{6}\right)a = 1a = a$

49. Applying the distributive property: $8(x+2) = 8 \bullet x + 8 \bullet 2 = 8x+16$
51. Applying the distributive property: $8(x-2) = 8 \bullet x - 8 \bullet 2 = 8x-16$
53. Applying the distributive property: $4(y+1) = 4 \bullet y + 4 \bullet 1 = 4y+4$
55. Applying the distributive property: $3(6x+5) = 3 \bullet 6x + 3 \bullet 5 = 18x+15$
57. Applying the distributive property: $2(3a+7) = 2 \bullet 3a + 2 \bullet 7 = 6a+14$
59. Applying the distributive property: $9(6y-8) = 9 \bullet 6y - 9 \bullet 8 = 54y-72$
61. Applying the distributive property: $\dfrac{1}{3}(3x+6) = \dfrac{1}{3} \bullet 3x + \dfrac{1}{3} \bullet 6 = x+2$
63. Applying the distributive property: $6(2x+3y) = 6 \bullet 2x + 6 \bullet 3y = 12x+18y$

65. Applying the distributive property: $4(3a - 2b) = 4 \cdot 3a - 4 \cdot 2b = 12a - 8b$

67. Applying the distributive property: $\dfrac{1}{2}(6x + 4y) = \dfrac{1}{2} \cdot 6x + \dfrac{1}{2} \cdot 4y = 3x + 2y$

69. Applying the distributive property: $4(a + 4) + 9 = 4a + 16 + 9 = 4a + 25$

71. Applying the distributive property: $2(3x + 5) + 2 = 6x + 10 + 2 = 6x + 12$

73. Applying the distributive property: $7(2x + 4) + 10 = 14x + 28 + 10 = 14x + 38$

75. Applying the distributive property: $\dfrac{1}{2}(4x + 2) = \dfrac{1}{2} \cdot 4x + \dfrac{1}{2} \cdot 2 = 2x + 1$

77. Applying the distributive property: $\dfrac{3}{4}(8x - 4) = \dfrac{3}{4} \cdot 8x - \dfrac{3}{4} \cdot 4 = 6x - 3$

79. Applying the distributive property: $\dfrac{5}{6}(6x + 12) = \dfrac{5}{6} \cdot 6x + \dfrac{5}{6} \cdot 12 = 5x + 10$

81. Applying the distributive property: $10\left(\dfrac{3}{5}x + \dfrac{1}{2}\right) = 10 \cdot \dfrac{3}{5}x + 10 \cdot \dfrac{1}{2} = 6x + 5$

83. Applying the distributive property: $15\left(\dfrac{1}{3}x + \dfrac{2}{5}\right) = 15 \cdot \dfrac{1}{3}x + 15 \cdot \dfrac{2}{5} = 5x + 6$

85. Applying the distributive property: $12\left(\dfrac{1}{2}m - \dfrac{5}{12}\right) = 12 \cdot \dfrac{1}{2}m - 12 \cdot \dfrac{5}{12} = 6m - 5$

87. Applying the distributive property: $21\left(\dfrac{1}{3} + \dfrac{1}{7}x\right) = 21 \cdot \dfrac{1}{3} + 21 \cdot \dfrac{1}{7}x = 7 + 3x$

89. Applying the distributive property: $6\left(\dfrac{1}{2}x - \dfrac{1}{3}y\right) = 6 \cdot \dfrac{1}{2}x - 6 \cdot \dfrac{1}{3}y = 3x - 2y$

91. Applying the distributive property: $0.09(x + 2,000) = 0.09x + 180$

93. Applying the distributive property: $0.12(x + 500) = 0.12x + 60$

95. Applying the distributive property: $a\left(1 + \dfrac{1}{a}\right) = a \cdot 1 + a \cdot \dfrac{1}{a} = a + 1$

97. Applying the distributive property: $a\left(\dfrac{1}{a} - 1\right) = a \cdot \dfrac{1}{a} - a \cdot 1 = 1 - a$

99. No. The man cannot reverse the order of putting on his socks and putting on his shoes.

101. Division is not a commutative operation. For example, $8 \div 4 = 2$ while $4 \div 8 = \dfrac{1}{2}$.

103. Computing his hours worked:

$$4(2 + 3) = 4(5) = 20 \text{ hours} \qquad\qquad 4 \cdot 2 + 4 \cdot 3 = 8 + 12 = 20 \text{ hours}$$

1.6 Multiplication of Real Numbers

1.   Finding the product: $7(-6) = -42$ 3.   Finding the product: $-8(2) = -16$

5.   Finding the product: $-3(-1) = 3$ 7.   Finding the product: $-11(-11) = 121$

9.   Using order of operations: $-3(2)(-1) = 6$ 11.   Using order of operations: $-3(-4)(-5) = -60$

13.   Using order of operations: $-2(-4)(-3)(-1) = 24$

15.   Using order of operations: $(-7)^2 = (-7)(-7) = 49$

17.   Using order of operations: $(-3)^3 = (-3)(-3)(-3) = -27$

19.   Using order of operations: $-2(2-5) = -2(-3) = 6$

21.   Using order of operations: $-5(8-10) = -5(-2) = 10$

23.   Using order of operations: $(4-7)(6-9) = (-3)(-3) = 9$

25.   Using order of operations: $(-3-2)(-5-4) = (-5)(-9) = 45$

27.   Using order of operations: $-3(-6) + 4(-1) = 18 + (-4) = 14$

29.   Using order of operations: $2(3) - 3(-4) + 4(-5) = 6 + 12 + (-20) = 18 + (-20) = -2$

31.   Using order of operations: $4(-3)^2 + 5(-6)^2 = 4(9) + 5(36) = 36 + 180 = 216$

33.   Using order of operations: $7(-2)^3 - 2(-3)^3 = 7(-8) - 2(-27) = -56 + 54 = -2$

35.   Using order of operations: $6 - 4(8-2) = 6 - 4(6) = 6 - 24 = 6 + (-24) = -18$

37.   Using order of operations: $9 - 4(3-8) = 9 - 4(-5) = 9 + 20 = 29$

39.   Using order of operations: $-4(3-8) - 6(2-5) = -4(-5) - 6(-3) = 20 + 18 = 38$

41.   Using order of operations: $7 - 2[-6 - 4(-3)] = 7 - 2(-6 + 12) = 7 - 2(6) = 7 - 12 = 7 + (-12) = -5$

43.   Using order of operations:

$$7 - 3[2(-4-4) - 3(-1-1)] = 7 - 3[2(-8) - 3(-2)]$$
$$= 7 - 3(-16 + 6)$$
$$= 7 - 3(-10)$$
$$= 7 + 30$$
$$= 37$$

45.   Using order of operations:

$$8 - 6[-2(-3-1) + 4(-2-3)] = 8 - 6[-2(-4) + 4(-5)]$$
$$= 8 - 6[8 + (-20)]$$
$$= 8 - 6(-12)$$
$$= 8 + 72$$
$$= 80$$

47.   Multiplying the fractions: $-\dfrac{2}{3} \cdot \dfrac{5}{7} = -\dfrac{2 \cdot 5}{3 \cdot 7} = -\dfrac{10}{21}$

49.   Multiplying the fractions: $-8\left(\dfrac{1}{2}\right) = -\dfrac{8}{1} \cdot \dfrac{1}{2} = -\dfrac{8}{2} = -4$

51.   Multiplying the fractions: $-\dfrac{3}{4}\left(-\dfrac{4}{3}\right) = -\dfrac{3}{4} \cdot \left(-\dfrac{4}{3}\right) = \dfrac{12}{12} = 1$

53. Multiplying the fractions: $\left(-\dfrac{3}{4}\right)^2 = \left(-\dfrac{3}{4}\right)\left(-\dfrac{3}{4}\right) = \dfrac{9}{16}$

55. Multiplying the expressions: $-2(4x) = (-2 \bullet 4)x = -8x$

57. Multiplying the expressions: $-7(-6x) = [-7 \bullet (-6)]x = 42x$

59. Multiplying the expressions: $-\dfrac{1}{3}(-3x) = \left[-\dfrac{1}{3} \bullet (-3)\right]x = 1x = x$

61. Simplifying the expression: $-4(a+2) = -4a + (-4)(2) = -4a - 8$

63. Simplifying the expression: $-\dfrac{1}{2}(3x-6) = -\dfrac{3}{2}x - \dfrac{1}{2}(-6) = -\dfrac{3}{2}x + 3$

65. Simplifying the expression: $-3(2x-5) - 7 = -6x + 15 - 7 = -6x + 8$

67. Simplifying the expression: $-5(3x+4) - 10 = -15x - 20 - 10 = -15x - 30$

69. Writing the expression: $3(-10) + 5 = -30 + 5 = -25$

71. Writing the expression: $2(-4x) = -8x$

73. Writing the expression: $-9 \bullet 2 - 8 = -18 + (-8) = -26$

75. The pattern is to multiply by 2, so the next number is $4 \bullet 2 = 8$.

77. The pattern is to multiply by –2, so the next number is $40 \bullet (-2) = -80$.

79. The pattern is to multiply by $\dfrac{1}{2}$, so the next number is $\dfrac{1}{4} \bullet \dfrac{1}{2} = \dfrac{1}{8}$.

81. The pattern is to multiply by –2, so the next number is $12 \bullet (-2) = -24$.

83. Simplifying the expression: $3(x-5) + 4 = 3x - 15 + 4 = 3x - 11$

85. Simplifying the expression: $2(3) - 4 - 3(-4) = 6 - 4 + 12 = 2 + 12 = 14$

87. Simplifying the expression: $\left(\dfrac{1}{2} \bullet 18\right)^2 = (9)^2 = 9 \bullet 9 = 81$

89. Simplifying the expression: $\left(\dfrac{1}{2} \bullet 3\right)^2 = \left(\dfrac{3}{2}\right)^2 = \left(\dfrac{3}{2}\right)\left(\dfrac{3}{2}\right) = \dfrac{9}{4}$

91. Simplifying the expression: $-\dfrac{1}{3}(-2x+6) = -\dfrac{1}{3}(-2x) - \dfrac{1}{3}(6) = \dfrac{2}{3}x - 2$

93. Simplifying the expression: $8\left(-\dfrac{1}{4}x + \dfrac{1}{8}y\right) = 8\left(-\dfrac{1}{4}x\right) + 8\left(\dfrac{1}{8}y\right) = -2x + y$

95. The temperature is: $25° - 4(6°) = 25° - 24° = 1°$

97. The net change in calories is: $2(630) - 3(265) = 1260 - 795 = 465$ calories

1.7 Division of Real Numbers

1. Finding the quotient: $\dfrac{8}{-4} = -2$

3. Finding the quotient: $\dfrac{-48}{16} = -3$

5. Finding the quotient: $\dfrac{-7}{21} = -\dfrac{1}{3}$

7. Finding the quotient: $\dfrac{-39}{-13} = 3$

9. Finding the quotient: $\dfrac{-6}{-42} = \dfrac{1}{7}$

11. Finding the quotient: $\dfrac{0}{-32} = 0$

13. Performing the operations: $-3 + 12 = 9$

15. Performing the operations: $-3 - 12 = -3 + (-12) = -15$

17. Performing the operations: $-3(12) = -36$

19. Performing the operations: $-3 \div 12 = \dfrac{-3}{12} = -\dfrac{1}{4}$

21. Dividing and reducing: $\dfrac{4}{5} \div \dfrac{3}{4} = \dfrac{4}{5} \bullet \dfrac{4}{3} = \dfrac{16}{15}$

23. Dividing and reducing: $-\dfrac{5}{6} \div \left(-\dfrac{5}{8}\right) = -\dfrac{5}{6} \bullet \left(-\dfrac{8}{5}\right) = \dfrac{40}{30} = \dfrac{4}{3}$

25. Dividing and reducing: $\dfrac{10}{13} \div \left(-\dfrac{5}{4}\right) = \dfrac{10}{13} \bullet \left(-\dfrac{4}{5}\right) = -\dfrac{40}{65} = -\dfrac{8}{13}$

27. Dividing and reducing: $-\dfrac{5}{6} \div \dfrac{5}{6} = -\dfrac{5}{6} \bullet \dfrac{6}{5} = -\dfrac{30}{30} = -1$

29. Dividing and reducing: $-\dfrac{3}{4} \div \left(-\dfrac{3}{4}\right) = -\dfrac{3}{4} \bullet \left(-\dfrac{4}{3}\right) = \dfrac{12}{12} = 1$

31. Using order of operations: $\dfrac{3(-2)}{-10} = \dfrac{-6}{-10} = \dfrac{3}{5}$

33. Using order of operations: $\dfrac{-5(-5)}{-15} = \dfrac{25}{-15} = -\dfrac{5}{3}$

35. Using order of operations: $\dfrac{-8(-7)}{-28} = \dfrac{56}{-28} = -2$

37. Using order of operations: $\dfrac{27}{4-13} = \dfrac{27}{-9} = -3$

39. Using order of operations: $\dfrac{20-6}{5-5} = \dfrac{14}{0}$, which is undefined

41. Using order of operations: $\dfrac{-3+9}{2 \bullet 5 - 10} = \dfrac{6}{10-10} = \dfrac{6}{0}$, which is undefined

43. Using order of operations: $\dfrac{15(-5)-25}{2(-10)} = \dfrac{-75-25}{-20} = \dfrac{-100}{-20} = 5$

45. Using order of operations: $\dfrac{27-2(-4)}{-3(5)} = \dfrac{27+8}{-15} = \dfrac{35}{-15} = -\dfrac{7}{3}$

47. Using order of operations: $\dfrac{12-6(-2)}{12(-2)} = \dfrac{12+12}{-24} = \dfrac{24}{-24} = -1$

49. Using order of operations: $\dfrac{5^2-2^2}{-5+2}=\dfrac{25-4}{-3}=\dfrac{21}{-3}=-7$

51. Using order of operations: $\dfrac{8^2-2^2}{8^2+2^2}=\dfrac{64-4}{64+4}=\dfrac{60}{68}=\dfrac{15}{17}$

53. Using order of operations: $\dfrac{(5+3)^2}{-5^2-3^2}=\dfrac{8^2}{-25-9}=\dfrac{64}{-34}=-\dfrac{32}{17}$

55. Using order of operations: $\dfrac{(8-4)^2}{8^2-4^2}=\dfrac{4^2}{64-16}=\dfrac{16}{48}=\dfrac{1}{3}$

57. Using order of operations: $\dfrac{-4\bullet 3^2-5\bullet 2^2}{-8(7)}=\dfrac{-4\bullet 9-5\bullet 4}{-56}=\dfrac{-36-20}{-56}=\dfrac{-56}{-56}=1$

59. Using order of operations: $\dfrac{3\bullet 10^2+4\bullet 10+5}{345}=\dfrac{300+40+5}{345}=\dfrac{345}{345}=1$

61. Using order of operations: $\dfrac{7-[(2-3)-4]}{-1-2-3}=\dfrac{7-(-1-4)}{-6}=\dfrac{7-(-5)}{-6}=\dfrac{7+5}{-6}=\dfrac{12}{-6}=-2$

63. Using order of operations: $\dfrac{6(-4)-2(5-8)}{-6-3-5}=\dfrac{-24-2(-3)}{-14}=\dfrac{-24+6}{-14}=\dfrac{-18}{-14}=\dfrac{9}{7}$

65. Using order of operations: $\dfrac{3(-5-3)+4(7-9)}{5(-2)+3(-4)}=\dfrac{3(-8)+4(-2)}{-10+(-12)}=\dfrac{-24+(-8)}{-22}=\dfrac{-32}{-22}=\dfrac{16}{11}$

67. Using order of operations: $\dfrac{|3-9|}{3-9}=\dfrac{|-6|}{-6}=\dfrac{6}{-6}=-1$

69. **a.** Simplifying: $20\div 4\bullet 5=5\bullet 5=25$ **b.** Simplifying: $-20\div 4\bullet 5=-5\bullet 5=-25$
   c. Simplifying: $20\div(-4)\bullet 5=-5\bullet 5=-25$ **d.** Simplifying: $20\div 4(-5)=5(-5)=-25$
   e. Simplifying: $-20\div 4(-5)=-5(-5)=25$

71. **a.** Simplifying: $8\div\dfrac{4}{5}=8\bullet\dfrac{5}{4}=10$

   b. Simplifying: $8\div\dfrac{4}{5}-10=8\bullet\dfrac{5}{4}-10=10-10=0$

   c. Simplifying: $8\div\dfrac{4}{5}(-10)=8\bullet\dfrac{5}{4}(-10)=10(-10)=-100$

   d. Simplifying: $8\div\left(-\dfrac{4}{5}\right)-10=8\bullet\left(-\dfrac{5}{4}\right)-10=-10-10=-20$

73. Applying the distributive property: $10\left(\dfrac{x}{2}+\dfrac{3}{5}\right)=10\left(\dfrac{x}{2}\right)+10\left(\dfrac{3}{5}\right)=5x+6$

75. Applying the distributive property: $15\left(\dfrac{x}{5}+\dfrac{4}{3}\right)=15\left(\dfrac{x}{5}\right)+15\left(\dfrac{4}{3}\right)=3x+20$

77. Applying the distributive property: $x\left(\dfrac{3}{x}+1\right)=x\left(\dfrac{3}{x}\right)+x(1)=3+x$

79. Applying the distributive property: $21\left(\dfrac{x}{7}-\dfrac{y}{3}\right)=21\left(\dfrac{x}{7}\right)-21\left(\dfrac{y}{3}\right)=3x-7y$

81. Applying the distributive property: $a\left(\dfrac{3}{a}-\dfrac{2}{a}\right)=a\left(\dfrac{3}{a}\right)-a\left(\dfrac{2}{a}\right)=3-2=1$

83. The quotient is $\dfrac{-12}{-4}=3$.

85. The number is –10, since $\dfrac{-10}{-5}=2$.

87. The number is –3, since $\dfrac{27}{-3}=-9$.

89. The expression is: $\dfrac{-20}{4}-3=-5-3=-8$

91. Each person would lose: $\dfrac{13600-15000}{4}=\dfrac{-1400}{4}=-350=\$350$ loss

93. The change per hour is: $\dfrac{61°-75°}{4}=\dfrac{-14°}{4}=-3.5°$ per hour

95. Dividing: $\dfrac{12\text{ yards}}{\sqrt[6]{7}\text{ yards/blanket}}=\dfrac{12}{1}\cdot\dfrac{7}{6}=14$ blankets

97. Dividing: $\dfrac{12\text{ pounds}}{\sqrt[1]{4}\text{ pounds/bag}}=\dfrac{12}{1}\cdot\dfrac{4}{1}=48$ bags

99. Dividing: $\dfrac{\sqrt[3]{4}\text{ teaspoon}}{\sqrt[1]{8}\text{ teaspoon}}=\dfrac{3}{4}\cdot\dfrac{8}{1}=6$ eighth-teaspoons

101. Dividing: $\dfrac{14\text{ pints}}{\sqrt[1]{2}\text{ pints/carton}}=\dfrac{14}{1}\cdot\dfrac{2}{1}=28$ cartons

103. **a.**   Since they predict $50 revenue for every 25 people, their projected revenue is:

$\dfrac{\$50}{25\text{ people}}\cdot10,000\text{ people}=\$20,000$

   b.   Since they predict $50 revenue for every 25 people, their projected revenue is:

$\dfrac{\$50}{25\text{ people}}\cdot25,000\text{ people}=\$50,000$

   c.   Since they predict $50 revenue for every 25 people, their projected revenue is:

$\dfrac{\$50}{25\text{ people}}\cdot5,000\text{ people}=\$10,000$

Since this projected revenue is more than the $5,000 cost for the list, it is a wise purchase.

1.8 Subsets of the Real Numbers

1.   The whole numbers are: 0, 1

3.   The rational numbers are: $-3, -2.5, 0, 1, \dfrac{3}{2}$

5.   The real numbers arc: $-3, -2.5, 0, 1, \dfrac{3}{2}, \sqrt{15}$

7.   The integers are: $-10, -8, -2, 9$

9.   The irrational numbers are: π

11.   true

13.   false

15.   false

17.   true

19.   This number is composite: $48 = 6 \bullet 8 = (2 \bullet 3) \bullet (2 \bullet 2 \bullet 2) = 2^4 \bullet 3$

21.   This number is prime.

23.   This number is composite: $1023 = 3 \bullet 341 = 3 \bullet 11 \bullet 31$

25.   Factoring the number: $144 = 12 \bullet 12 = (3 \bullet 4) \bullet (3 \bullet 4) = (3 \bullet 2 \bullet 2) \bullet (3 \bullet 2 \bullet 2) = 2^4 \bullet 3^2$

27.   Factoring the number: $38 = 2 \bullet 19$

29.   Factoring the number: $105 = 5 \bullet 21 = 5 \bullet (3 \bullet 7) = 3 \bullet 5 \bullet 7$

31.   Factoring the number: $180 = 10 \bullet 18 = (2 \bullet 5) \bullet (3 \bullet 6) = (2 \bullet 5) \bullet (3 \bullet 2 \bullet 3) = 2^2 \bullet 3^2 \bullet 5$

33.   Factoring the number: $385 = 5 \bullet 77 = 5 \bullet (7 \bullet 11) = 5 \bullet 7 \bullet 11$

35.   Factoring the number: $121 = 11 \bullet 11 = 11^2$

37.   Factoring the number: $420 = 10 \bullet 42 = (2 \bullet 5) \bullet (7 \bullet 6) = (2 \bullet 5) \bullet (7 \bullet 2 \bullet 3) = 2^2 \bullet 3 \bullet 5 \bullet 7$

39.   Factoring the number: $620 = 10 \bullet 62 = (2 \bullet 5) \bullet (2 \bullet 31) = 2^2 \bullet 5 \bullet 31$

41.   Reducing the fraction: $\dfrac{105}{165} = \dfrac{3 \bullet 5 \bullet 7}{3 \bullet 5 \bullet 11} = \dfrac{7}{11}$

43.   Reducing the fraction: $\dfrac{525}{735} = \dfrac{3 \bullet 5 \bullet 5 \bullet 7}{3 \bullet 5 \bullet 7 \bullet 7} = \dfrac{5}{7}$

45.   Reducing the fraction: $\dfrac{385}{455} = \dfrac{5 \bullet 7 \bullet 11}{5 \bullet 7 \bullet 13} = \dfrac{11}{13}$

47.   Reducing the fraction: $\dfrac{322}{345} = \dfrac{2 \bullet 7 \bullet 23}{3 \bullet 5 \bullet 23} = \dfrac{2 \bullet 7}{3 \bullet 5} = \dfrac{14}{15}$

49.   Reducing the fraction: $\dfrac{205}{369} = \dfrac{5 \bullet 41}{3 \bullet 3 \bullet 41} = \dfrac{5}{3 \bullet 3} = \dfrac{5}{9}$

51.   Reducing the fraction: $\dfrac{215}{344} = \dfrac{5 \bullet 43}{2 \bullet 2 \bullet 2 \bullet 43} = \dfrac{5}{2 \bullet 2 \bullet 2} = \dfrac{5}{8}$

53.   Factoring into prime numbers: $6^3 = (2 \bullet 3)^3 = 2^3 \bullet 3^3$

55.   Factoring into prime numbers:

$$9^4 \bullet 16^2 = (3 \bullet 3)^4 \bullet (2 \bullet 2 \bullet 2 \bullet 2)^2 = 3^4 \bullet 3^4 \bullet 2^2 \bullet 2^2 \bullet 2^2 \bullet 2^2 = 2^8 \bullet 3^8$$

57.   Simplifying and factoring:

$$3 \bullet 8 + 3 \bullet 7 + 3 \bullet 5 = 24 + 21 + 15 = 60 = 6 \bullet 10 = (2 \bullet 3) \bullet (2 \bullet 5) = 2^2 \bullet 3 \bullet 5$$

59.   They are not a subset of the irrational numbers.

61.   8, 21, and 34 are Fibonacci numbers that are composite numbers.

1.9 Addition and Subtraction with Fractions

1. Combining the fractions: $\dfrac{3}{6} + \dfrac{1}{6} = \dfrac{4}{6} = \dfrac{2}{3}$

3. Combining the fractions: $\dfrac{3}{8} - \dfrac{5}{8} = -\dfrac{2}{8} = -\dfrac{1}{4}$

5. Combining the fractions: $-\dfrac{1}{4} + \dfrac{3}{4} = \dfrac{2}{4} = \dfrac{1}{2}$

7. Combining the fractions: $\dfrac{x}{3} - \dfrac{1}{3} = \dfrac{x-1}{3}$

9. Combining the fractions: $\dfrac{1}{4} + \dfrac{2}{4} + \dfrac{3}{4} = \dfrac{6}{4} = \dfrac{3}{2}$

11. Combining the fractions: $\dfrac{x+7}{2} - \dfrac{1}{2} = \dfrac{x+7-1}{2} = \dfrac{x+6}{2}$

13. Combining the fractions: $\dfrac{1}{10} - \dfrac{3}{10} - \dfrac{4}{10} = -\dfrac{6}{10} = -\dfrac{3}{5}$

15. Combining the fractions: $\dfrac{1}{a} + \dfrac{4}{a} + \dfrac{5}{a} = \dfrac{10}{a}$

17. Completing the table:

First Number a	Second Number b	The Sum of a and b $a+b$
$\dfrac{1}{2}$	$\dfrac{1}{3}$	$\dfrac{5}{6}$
$\dfrac{1}{3}$	$\dfrac{1}{4}$	$\dfrac{7}{12}$
$\dfrac{1}{4}$	$\dfrac{1}{5}$	$\dfrac{9}{20}$
$\dfrac{1}{5}$	$\dfrac{1}{6}$	$\dfrac{11}{30}$

19. Completing the table:

First Number a	Second Number b	The Sum of a and b $a+b$
$\dfrac{1}{12}$	$\dfrac{1}{2}$	$\dfrac{7}{12}$
$\dfrac{1}{12}$	$\dfrac{1}{3}$	$\dfrac{5}{12}$
$\dfrac{1}{12}$	$\dfrac{1}{4}$	$\dfrac{1}{3}$
$\dfrac{1}{12}$	$\dfrac{1}{6}$	$\dfrac{1}{4}$

21. Combining the fractions: $\dfrac{4}{9} + \dfrac{1}{3} = \dfrac{4}{9} + \dfrac{1 \cdot 3}{3 \cdot 3} = \dfrac{4}{9} + \dfrac{3}{9} = \dfrac{7}{9}$

23. Combining the fractions: $2 + \dfrac{1}{3} = \dfrac{2 \cdot 3}{1 \cdot 3} + \dfrac{1}{3} = \dfrac{6}{3} + \dfrac{1}{3} = \dfrac{7}{3}$

25. Combining the fractions: $-\dfrac{3}{4} + 1 = -\dfrac{3}{4} + \dfrac{1 \cdot 4}{1 \cdot 4} = -\dfrac{3}{4} + \dfrac{4}{4} = \dfrac{1}{4}$

27. Combining the fractions: $\dfrac{1}{2}+\dfrac{2}{3}=\dfrac{1\cdot 3}{2\cdot 3}+\dfrac{2\cdot 2}{3\cdot 2}=\dfrac{3}{6}+\dfrac{4}{6}=\dfrac{7}{6}$

29. Combining the fractions: $\dfrac{5}{12}-\left(-\dfrac{3}{8}\right)=\dfrac{5}{12}+\dfrac{3}{8}=\dfrac{5\cdot 2}{12\cdot 2}+\dfrac{3\cdot 3}{8\cdot 3}=\dfrac{10}{24}+\dfrac{9}{24}=\dfrac{19}{24}$

31. Combining the fractions: $-\dfrac{1}{20}+\dfrac{8}{30}=-\dfrac{1\cdot 3}{20\cdot 3}+\dfrac{8\cdot 2}{30\cdot 2}=-\dfrac{3}{60}+\dfrac{16}{60}=\dfrac{13}{60}$

33. First factor the denominators to find the LCM:

$\quad\quad 30=2\cdot 3\cdot 5$

$\quad\quad 42=2\cdot 3\cdot 7$

$\quad\quad LCM=2\cdot 3\cdot 5\cdot 7=210$

Combining the fractions: $\dfrac{17}{30}+\dfrac{11}{42}=\dfrac{17\cdot 7}{30\cdot 7}+\dfrac{11\cdot 5}{42\cdot 5}=\dfrac{119}{210}+\dfrac{55}{210}=\dfrac{174}{210}=\dfrac{2\cdot 3\cdot 29}{2\cdot 3\cdot 5\cdot 7}=\dfrac{29}{5\cdot 7}=\dfrac{29}{35}$

35. First factor the denominators to find the LCM:

$\quad\quad 84=2\cdot 2\cdot 3\cdot 7$

$\quad\quad 90=2\cdot 3\cdot 3\cdot 5$

$\quad\quad LCM=2\cdot 2\cdot 3\cdot 3\cdot 5\cdot 7=1260$

Combining the fractions: $\dfrac{25}{84}+\dfrac{41}{90}=\dfrac{25\cdot 15}{84\cdot 15}+\dfrac{41\cdot 14}{90\cdot 14}=\dfrac{375}{1,260}+\dfrac{574}{1,260}=\dfrac{949}{1,260}$

37. First factor the denominators to find the LCM:

$\quad\quad 126=2\cdot 3\cdot 3\cdot 7$

$\quad\quad 180=2\cdot 2\cdot 3\cdot 3\cdot 5$

$\quad\quad LCM=2\cdot 2\cdot 3\cdot 3\cdot 5\cdot 7=1260$

Combining the fractions:

$$\dfrac{13}{126}-\dfrac{13}{180}=\dfrac{13\cdot 10}{126\cdot 10}-\dfrac{13\cdot 7}{180\cdot 7}$$

$$=\dfrac{130}{1260}-\dfrac{91}{1260}$$

$$=\dfrac{39}{1260}$$

$$=\dfrac{3\cdot 13}{2\cdot 2\cdot 3\cdot 3\cdot 5\cdot 7}$$

$$=\dfrac{13}{2\cdot 2\cdot 3\cdot 5\cdot 7}$$

$$=\dfrac{13}{420}$$

39. Combining the fractions: $\dfrac{3}{4}+\dfrac{1}{8}+\dfrac{5}{6}=\dfrac{3\cdot 6}{4\cdot 6}+\dfrac{1\cdot 3}{8\cdot 3}+\dfrac{5\cdot 4}{6\cdot 4}=\dfrac{18}{24}+\dfrac{3}{24}+\dfrac{20}{24}=\dfrac{41}{24}$

41. Combining the fractions:

$$\dfrac{1}{2}+\dfrac{1}{3}+\dfrac{1}{4}+\dfrac{1}{6}=\dfrac{1\cdot 6}{2\cdot 6}+\dfrac{1\cdot 4}{3\cdot 4}+\dfrac{1\cdot 3}{4\cdot 3}+\dfrac{1\cdot 2}{6\cdot 2}=\dfrac{6}{12}+\dfrac{4}{12}+\dfrac{3}{12}+\dfrac{2}{12}=\dfrac{15}{12}=\dfrac{5}{4}$$

43. Combining the fractions: $1-\dfrac{5}{2}=1\cdot\dfrac{2}{2}-\dfrac{5}{2}=\dfrac{2}{2}-\dfrac{5}{2}=-\dfrac{3}{2}$

45. Combining the fractions: $1+\dfrac{1}{2}=1\cdot\dfrac{2}{2}+\dfrac{1}{2}=\dfrac{2}{2}+\dfrac{1}{2}=\dfrac{3}{2}$

47. The sum is given by: $\dfrac{3}{7}+2+\dfrac{1}{9}=\dfrac{3\cdot9}{7\cdot9}+\dfrac{2\cdot63}{1\cdot63}+\dfrac{1\cdot7}{9\cdot7}=\dfrac{27}{63}+\dfrac{126}{63}+\dfrac{7}{63}=\dfrac{160}{63}$

49. The difference is given by: $\dfrac{7}{8}-\dfrac{1}{4}=\dfrac{7}{8}-\dfrac{1\cdot2}{4\cdot2}=\dfrac{7}{8}-\dfrac{2}{8}=\dfrac{5}{8}$

51. The pattern is to add $-\dfrac{1}{3}$, so the fourth term is: $-\dfrac{1}{3}+\left(-\dfrac{1}{3}\right)=-\dfrac{2}{3}$

53. The pattern is to add $\dfrac{2}{3}$, so the fourth term is: $\dfrac{5}{3}+\dfrac{2}{3}=\dfrac{7}{3}$

55. The pattern is to multiply by $\dfrac{1}{5}$, so the fourth term is: $\dfrac{1}{25}\cdot\dfrac{1}{5}=\dfrac{1}{125}$

57. The perimeter is: $\dfrac{3}{8}+\dfrac{3}{8}+\dfrac{3}{8}+\dfrac{3}{8}=\dfrac{12}{8}=\dfrac{3}{2}=1\dfrac{1}{2}$ feet

59. The perimeter is: $\dfrac{4}{5}+\dfrac{3}{10}+\dfrac{4}{5}+\dfrac{3}{10}=\dfrac{8}{10}+\dfrac{3}{10}+\dfrac{8}{10}+\dfrac{3}{10}=\dfrac{22}{10}=\dfrac{11}{5}=2\dfrac{1}{5}$ centimeters (cm)

61. Adding: $\dfrac{1}{2}+4=\dfrac{1}{2}+\dfrac{8}{2}=\dfrac{9}{2}=4\dfrac{1}{2}$ pints

63. Multiplying: $\dfrac{5}{8}\cdot2,120=\dfrac{5}{8}\cdot\dfrac{2120}{1}=\$1,325$

65. Adding: $\dfrac{1}{4}+\dfrac{3}{20}=\dfrac{5}{20}+\dfrac{3}{20}=\dfrac{8}{20}=\dfrac{2}{5}$ of the students

67. Completing the table:

Grade	Number of Students	Fraction of Students
A	5	$\dfrac{1}{8}$
B	8	$\dfrac{1}{5}$
C	20	$\dfrac{1}{2}$
below C	7	$\dfrac{7}{40}$
Total	40	1

69. Dividing: $\dfrac{6\text{ acres}}{\frac{3}{5}\text{ acres/lot}}=\dfrac{6}{1}\cdot\dfrac{5}{3}=10$ lots

Chapter 1 Test

See www.mathtv.com for video solutions to all problems in this chapter test.

Chapter 2
Linear Equations and Inequalities

2.1 Simplifying Expressions

1. Simplifying the expression: $3x - 6x = (3 - 6)x = -3x$
3. Simplifying the expression: $-2a + a = (-2 + 1)a = -a$
5. Simplifying the expression: $7x + 3x + 2x = (7 + 3 + 2)x = 12x$
7. Simplifying the expression: $3a - 2a + 5a = (3 - 2 + 5)a = 6a$
9. Simplifying the expression: $4x - 3 + 2x = 4x + 2x - 3 = 6x - 3$
11. Simplifying the expression: $3a + 4a + 5 = 7a + 5$
13. Simplifying the expression: $2x - 3 + 3x - 2 = 2x + 3x - 3 - 2 = 5x - 5$
15. Simplifying the expression: $3a - 1 + a + 3 = 3a + a - 1 + 3 = 4a + 2$
17. Simplifying the expression: $-4x + 8 - 5x - 10 = -4x - 5x + 8 - 10 = -9x - 2$
19. Simplifying the expression: $7a + 3 + 2a + 3a = 7a + 2a + 3a + 3 = 12a + 3$
21. Simplifying the expression: $5(2x - 1) + 4 = 10x - 5 + 4 = 10x - 1$
23. Simplifying the expression: $7(3y + 2) - 8 = 21y + 14 - 8 = 21y + 6$
25. Simplifying the expression: $-3(2x - 1) + 5 = -6x + 3 + 5 = -6x + 8$
27. Simplifying the expression: $5 - 2(a + 1) = 5 - 2a - 2 = -2a - 2 + 5 = -2a + 3$
29. Simplifying the expression: $6 - 4(x - 5) = 6 - 4x + 20 = -4x + 20 + 6 = -4x + 26$
31. Simplifying the expression: $-9 - 4(2 - y) + 1 = -9 - 8 + 4y + 1 = 4y + 1 - 9 - 8 = 4y - 16$
33. Simplifying the expression: $-6 + 2(2 - 3x) + 1 = -6 + 4 - 6x + 1 = -6x - 6 + 4 + 1 = -6x - 1$
35. Simplifying the expression: $(4x - 7) - (2x + 5) = 4x - 7 - 2x - 5 = 4x - 2x - 7 - 5 = 2x - 12$
37. Simplifying the expression: $8(2a + 4) - (6a - 1) = 16a + 32 - 6a + 1 = 16a - 6a + 32 + 1 = 10a + 33$
39. Simplifying the expression: $3(x - 2) + (x - 3) = 3x - 6 + x - 3 = 3x + x - 6 - 3 = 4x - 9$
41. Simplifying the expression: $4(2y - 8) - (y + 7) = 8y - 32 - y - 7 = 8y - y - 32 - 7 = 7y - 39$
43. Simplifying the expression: $-9(2x + 1) - (x + 5) = -18x - 9 - x - 5 = -18x - x - 9 - 5 = -19x - 14$
45. Evaluating when $x = 2$: $3x - 1 = 3(2) - 1 = 6 - 1 = 5$
47. Evaluating when $x = 2$: $-2x - 5 = -2(2) - 5 = -4 - 5 = -9$
49. Evaluating when $x = 2$: $x^2 - 8x + 16 = (2)^2 - 8(2) + 16 = 4 - 16 + 16 = 4$
51. Evaluating when $x = 2$: $(x - 4)^2 = (2 - 4)^2 = (-2)^2 = 4$
53. Evaluating when $x = -5$: $7x - 4 - x - 3 = 7(-5) - 4 - (-5) - 3 = -35 - 4 + 5 - 3 = -42 + 5 = -37$
    Now simplifying the expression: $7x - 4 - x - 3 = 7x - x - 4 - 3 = 6x - 7$
    Evaluating when $x = -5$: $6x - 7 = 6(-5) - 7 = -30 - 7 = -37$
    Note that the two values are the same.

55.   Evaluating when $x = -5$:
$$5(2x+1)+4 = 5[2(-5)+1]+4 = 5(-10+1)+4 = 5(-9)+4 = -45+4 = -41$$
Now simplifying the expression: $5(2x+1)+4 = 10x+5+4 = 10x+9$
Evaluating when $x = -5$: $10x+9 = 10(-5)+9 = -50+9 = -41$
Note that the two values are the same.

57.   Evaluating when $x = -3$ and $y = 5$: $x^2 - 2xy + y^2 = (-3)^2 - 2(-3)(5) + (5)^2 = 9 + 30 + 25 = 64$

59.   Evaluating when $x = -3$ and $y = 5$: $(x-y)^2 = (-3-5)^2 = (-8)^2 = 64$

61.   Evaluating when $x = -3$ and $y = 5$:
$$x^2 + 6xy + 9y^2 = (-3)^2 + 6(-3)(5) + 9(5)^2 = 9 - 90 + 225 = 144$$

63.   Evaluating when $x = -3$ and $y = 5$: $(x+3y)^2 = [-3+3(5)]^2 = (-3+15)^2 = (12)^2 = 144$

65.   Evaluating when $x = \dfrac{1}{2}$: $12x - 3 = 12\left(\dfrac{1}{2}\right) - 3 = 6 - 3 = 3$

67.   Evaluating when $x = \dfrac{1}{4}$: $12x - 3 = 12\left(\dfrac{1}{4}\right) - 3 = 3 - 3 = 0$

69.   Evaluating when $x = \dfrac{3}{2}$: $12x - 3 = 12\left(\dfrac{3}{2}\right) - 3 = 18 - 3 = 15$

71.   Evaluating when $x = \dfrac{3}{4}$: $12x - 3 = 12\left(\dfrac{3}{4}\right) - 3 = 9 - 3 = 6$

73.   a.   Substituting the values for n:

n	1	2	3	4
$3n$	3	6	9	12

b.   Substituting the values for n:

n	1	2	3	4
n^3	1	8	27	64

75.   Substituting $n = 1, 2, 3, 4$:
$n = 1$: $3(1) - 2 = 3 - 2 = 1$
$n = 2$: $3(2) - 2 = 6 - 2 = 4$
$n = 3$: $3(3) - 2 = 9 - 2 = 7$
$n = 4$: $3(4) - 2 = 12 - 2 = 10$
The sequence is 1, 4, 7, 10, ..., which is an arithmetic sequence.

77.   Substituting $n = 1, 2, 3, 4$:
$n = 1$: $(1)^2 - 2(1) + 1 = 1 - 2 + 1 = 0$
$n = 2$: $(2)^2 - 2(2) + 1 = 4 - 4 + 1 = 1$
$n = 3$: $(3)^2 - 2(3) + 1 = 9 - 6 + 1 = 4$
$n = 4$: $(4)^2 - 2(4) + 1 = 16 - 8 + 1 = 9$
The sequence is 0, 1, 4, 9, ..., which is a sequence of squares.

79.   Simplifying: $7 - 3(2y+1) = 7 - 6y - 3 = -6y + 4$

81.   Simplifying: $0.08x + 0.09x = 0.17x$

83.   Simplifying: $(x+y) + (x-y) = x + y + x - y = 2x$

85.   Simplifying: $3x + 2(x-2) = 3x + 2x - 4 = 5x - 4$

87. Simplifying: $4(x+1)+3(x-3)=4x+4+3x-9=7x-5$

89. Simplifying: $x+(x+3)(-3)=x-3x-9=-2x-9$

91. Simplifying: $3(4x-2)-(5x-8)=12x-6-5x+8=7x+2$

93. Simplifying: $-(3x+1)-(4x-7)=-3x-1-4x+7=-7x+6$

95. Simplifying: $(x+3y)+3(2x-y)=x+3y+6x-3y=7x$

97. Simplifying: $3(2x+3y)-2(3x+5y)=6x+9y-6x-10y=-y$

99. Simplifying: $-6\left(\dfrac{1}{2}x-\dfrac{1}{3}y\right)+12\left(\dfrac{1}{4}x+\dfrac{2}{3}y\right)=-3x+2y+3x+8y=10y$

101. Simplifying: $0.08x+0.09(x+2{,}000)=0.08x+0.09x+180=0.17x+180$

103. Simplifying: $0.10x+0.12(x+500)=0.10x+0.12x+60=0.22x+60$

105. Evaluating the expression: $b^2-4ac=(-5)^2-4(1)(-6)=25-(-24)=25+24=49$

107. Evaluating the expression: $b^2-4ac=(4)^2-4(2)(-3)=16-(-24)=16+24=40$

109.   **a.**   Substituting $x=8{,}000$: $-0.0035(8000)+70=42°F$

      b.   Substituting $x=12{,}000$: $-0.0035(12000)+70=28°F$

      c.   Substituting $x=24{,}000$: $-0.0035(24000)+70=-14°F$

111.   **a.**   Substituting $t=10$: $35+0.25(10)=\$37.50$

      b.   Substituting $t=20$: $35+0.25(20)=\$40.00$

      c.   Substituting $t=30$: $35+0.25(30)=\$42.50$

113. Simplifying: $17-5=12$

115. Simplifying: $2-5=-3$

117. Simplifying: $-2.4+(-7.3)=-9.7$

119. Simplifying: $-\dfrac{1}{2}+\left(-\dfrac{3}{4}\right)=-\dfrac{1}{2}\cdot\dfrac{2}{2}-\dfrac{3}{4}=-\dfrac{2}{4}-\dfrac{3}{4}=-\dfrac{5}{4}$

121. Simplifying: $4(2\cdot9-3)-7=4(18-3)-7=4(15)-7=60-7=53$

123. Simplifying: $4(2a-3)-7a=8a-12-7a=a-12$

125. Evaluating when $x=5$: $2(5)-3=10-3=7$

2.2 Addition Property of Equality

1. Solving the equation:
$$x - 3 = 8$$
$$x - 3 + 3 = 8 + 3$$
$$x = 11$$

3. Solving the equation:
$$x + 2 = 6$$
$$x + 2 + (-2) = 6 + (-2)$$
$$x = 4$$

5. Solving the equation:
$$a + \frac{1}{2} = -\frac{1}{4}$$
$$a + \frac{1}{2} + \left(-\frac{1}{2}\right) = -\frac{1}{4} + \left(-\frac{1}{2}\right)$$
$$a = -\frac{1}{4} + \left(-\frac{2}{4}\right)$$
$$a = -\frac{3}{4}$$

7. Solving the equation:
$$x + 2.3 = -3.5$$
$$x + 2.3 + (-2.3) = -3.5 + (-2.3)$$
$$x = -5.8$$

9. Solving the equation:
$$y + 11 = -6$$
$$y + 11 + (-11) = -6 + (-11)$$
$$y = -17$$

11. Solving the equation:
$$x - \frac{5}{8} = -\frac{3}{4}$$
$$x - \frac{5}{8} + \frac{5}{8} = -\frac{3}{4} + \frac{5}{8}$$
$$x = -\frac{6}{8} + \frac{5}{8}$$
$$x = -\frac{1}{8}$$

13. Solving the equation:
$$m - 6 = -10$$
$$m - 6 + 6 = -10 + 6$$
$$m = -4$$

15. Solving the equation:
$$6.9 + x = 3.3$$
$$-6.9 + 6.9 + x = -6.9 + 3.3$$
$$x = -3.6$$

17. Solving the equation:
$$5 = a + 4$$
$$5 + (-4) = a + 4 + (-4)$$
$$a = 1$$

19. Solving the equation:
$$-\frac{5}{9} = x - \frac{2}{5}$$
$$-\frac{5}{9} + \frac{2}{5} = x - \frac{2}{5} + \frac{2}{5}$$
$$-\frac{25}{45} + \frac{18}{45} = x$$
$$x = -\frac{7}{45}$$

21. Solving the equation:

$$4x + 2 - 3x = 4 + 1$$
$$x + 2 = 5$$
$$x + 2 + (-2) = 5 + (-2)$$
$$x = 3$$

23. Solving the equation:

$$8a - \frac{1}{2} - 7a = \frac{3}{4} + \frac{1}{8}$$
$$a - \frac{1}{2} = \frac{6}{8} + \frac{1}{8}$$
$$a - \frac{1}{2} = \frac{7}{8}$$
$$a - \frac{1}{2} + \frac{1}{2} = \frac{7}{8} + \frac{1}{2}$$
$$a = \frac{7}{8} + \frac{4}{8}$$
$$a = \frac{11}{8}$$

25. Solving the equation:
$$-3 - 4x + 5x = 18$$
$$-3 + x = 18$$
$$3 - 3 + x = 3 + 18$$
$$x = 21$$

27. Solving the equation:
$$-11x + 2 + 10x + 2x = 9$$
$$x + 2 = 9$$
$$x + 2 + (-2) = 9 + (-2)$$
$$x = 7$$

29. Solving the equation:
$$-2.5 + 4.8 = 8x - 1.2 - 7x$$
$$2.3 = x - 1.2$$
$$2.3 + 1.2 = x - 1.2 + 1.2$$
$$x = 3.5$$

31. Solving the equation:
$$2y - 10 + 3y - 4y = 18 - 6$$
$$y - 10 = 12$$
$$y - 10 + 10 = 12 + 10$$
$$y = 22$$

33. Solving the equation:
$$2(x + 3) - x = 4$$
$$2x + 6 - x = 4$$
$$x + 6 = 4$$
$$x + 6 + (-6) = 4 + (-6)$$
$$x = -2$$

35. Solving the equation:
$$-3(x - 4) + 4x = 3 - 7$$
$$-3x + 12 + 4x = -4$$
$$x + 12 = -4$$
$$x + 12 + (-12) = -4 + (-12)$$
$$x = -16$$

37. Solving the equation:
$$5(2a + 1) - 9a = 8 - 6$$
$$10a + 5 - 9a = 2$$
$$a + 5 = 2$$
$$a + 5 + (-5) = 2 + (-5)$$
$$a = -3$$

39. Solving the equation:
$$-(x + 3) + 2x - 1 = 6$$
$$-x - 3 + 2x - 1 = 6$$
$$x - 4 = 6$$
$$x - 4 + 4 = 6 + 4$$
$$x = 10$$

41. Solving the equation:
$$4y - 3(y - 6) + 2 = 8$$
$$4y - 3y + 18 + 2 = 8$$
$$y + 20 = 8$$
$$y + 20 + (-20) = 8 + (-20)$$
$$y = -12$$

43. Solving the equation:
$$-3(2m - 9) + 7(m - 4) = 12 - 9$$
$$-6m + 27 + 7m - 28 = 3$$
$$m - 1 = 3$$
$$m - 1 + 1 = 3 + 1$$
$$m = 4$$

45. Solving the equation:
$$4x = 3x + 2$$
$$4x + (-3x) = 3x + (-3x) + 2$$
$$x = 2$$

47. Solving the equation:
$$8a = 7a - 5$$
$$8a + (-7a) = 7a + (-7a) - 5$$
$$a = -5$$

49. Solving the equation:
$$2x = 3x + 1$$
$$(-2x) + 2x = (-2x) + 3x + 1$$
$$0 = x + 1$$
$$0 + (-1) = x + 1 + (-1)$$
$$x = -1$$

51. Solving the equation:
$$3y + 4 = 2y + 1$$
$$3y + (-2y) + 4 = 2y + (-2y) + 1$$
$$y + 4 = 1$$
$$y + 4 + (-4) = 1 + (-4)$$
$$y = -3$$

53. Solving the equation:
$$2m - 3 = m + 5$$
$$2m + (-m) - 3 = m + (-m) + 5$$
$$m - 3 = 5$$
$$m - 3 + 3 = 5 + 3$$
$$m = 8$$

55. Solving the equation:
$$4x - 7 = 5x + 1$$
$$4x + (-4x) - 7 = 5x + (-4x) + 1$$
$$-7 = x + 1$$
$$-7 + (-1) = x + 1 + (-1)$$
$$x = -8$$

57. Solving the equation:
$$5x - \frac{2}{3} = 4x + \frac{4}{3}$$
$$5x + (-4x) - \frac{2}{3} = 4x + (-4x) + \frac{4}{3}$$
$$x - \frac{2}{3} = \frac{4}{3}$$
$$x - \frac{2}{3} + \frac{2}{3} = \frac{4}{3} + \frac{2}{3}$$
$$x = \frac{6}{3} = 2$$

59. Solving the equation:
$$8a - 7.1 = 7a + 3.9$$
$$8a + (-7a) - 7.1 = 7a + (-7a) + 3.9$$
$$a - 7.1 = 3.9$$
$$a - 7.1 + 7.1 = 3.9 + 7.1$$
$$a = 11$$

61. Solving the equation:
$$11y - 2.9 = 12y + 2.9$$
$$11y + (-11y) - 2.9 = 12y + (-11y) + 2.9$$
$$-2.9 = y + 2.9$$
$$-2.9 - 2.9 = y + 2.9 - 2.9$$
$$y = -5.8$$

63. a. Solving for R:
$$T + R + A = 100$$
$$88 + R + 6 = 100$$
$$94 + R = 100$$
$$R = 6\%$$

b. Solving for R:
$$T + R + A = 100$$
$$0 + R + 95 = 100$$
$$95 + R = 100$$
$$R = 5\%$$

c.   Solving for A:
$$T + R + A = 100$$
$$0 + 98 + A = 100$$
$$98 + A = 100$$
$$A = 2\%$$

d.   Solving for R:
$$T + R + A = 100$$
$$0 + R + 25 = 100$$
$$25 + R = 100$$
$$R = 75\%$$

65. Simplifying: $\dfrac{3}{2}\left(\dfrac{2}{3}y\right) = y$

67. Simplifying: $\dfrac{1}{5}(5x) = x$

69. Simplifying: $\dfrac{1}{5}(30) = 6$

71. Simplifying: $\dfrac{3}{2}(4) = \dfrac{12}{2} = 6$

73. Simplifying: $12\left(-\dfrac{3}{4}\right) = -\dfrac{36}{4} = -9$

75. Simplifying: $\dfrac{3}{2}\left(-\dfrac{5}{4}\right) = -\dfrac{15}{8}$

77. Simplifying: $13 + (-5) = 8$

79. Simplifying: $-\dfrac{3}{4} + \left(-\dfrac{1}{2}\right) = -\dfrac{3}{4} - \dfrac{1}{2} \cdot \dfrac{2}{2} = -\dfrac{3}{4} - \dfrac{2}{4} = -\dfrac{5}{4}$

81. Simplifying: $7x + (-4x) = 3x$

2.3 Multiplication Property of Equality

1.   Solving the equation:
$$5x = 10$$
$$\dfrac{1}{5}(5x) = \dfrac{1}{5}(10)$$
$$x = 2$$

3.   Solving the equation:
$$7a = 28$$
$$\dfrac{1}{7}(7a) = \dfrac{1}{7}(28)$$
$$a = 4$$

5.   Solving the equation:
$$-8x = 4$$
$$-\dfrac{1}{8}(-8x) = -\dfrac{1}{8}(4)$$
$$x = -\dfrac{1}{2}$$

7.   Solving the equation:
$$8m = -16$$
$$\dfrac{1}{8}(8m) = \dfrac{1}{8}(-16)$$
$$m = -2$$

9.   Solving the equation:
$$-3x = -9$$
$$-\dfrac{1}{3}(-3x) = -\dfrac{1}{3}(-9)$$
$$x = 3$$

11. Solving the equation:
$$-7y = -28$$
$$-\dfrac{1}{7}(-7y) = -\dfrac{1}{7}(-28)$$
$$y = 4$$

13. Solving the equation:
$$2x = 0$$
$$\dfrac{1}{2}(2x) = \dfrac{1}{2}(0)$$
$$x = 0$$

15. Solving the equation:
$$-5x = 0$$
$$-\dfrac{1}{5}(-5x) = -\dfrac{1}{5}(0)$$
$$x = 0$$

17.   Solving the equation:

$$\frac{x}{3} = 2$$

$$3\left(\frac{x}{3}\right) = 3(2)$$

$$x = 6$$

19.   Solving the equation:

$$-\frac{m}{5} = 10$$

$$-5\left(-\frac{m}{5}\right) = -5(10)$$

$$m = -50$$

21.   Solving the equation:

$$-\frac{x}{2} = -\frac{3}{4}$$

$$-2\left(-\frac{x}{2}\right) = -2\left(-\frac{3}{4}\right)$$

$$x = \frac{3}{2}$$

23.   Solving the equation:

$$\frac{2}{3}a = 8$$

$$\frac{3}{2}\left(\frac{2}{3}a\right) = \frac{3}{2}(8)$$

$$a = 12$$

25.   Solving the equation:

$$-\frac{3}{5}x = \frac{9}{5}$$

$$-\frac{5}{3}\left(-\frac{3}{5}x\right) = -\frac{5}{3}\left(\frac{9}{5}\right)$$

$$x = -3$$

27.   Solving the equation:

$$-\frac{5}{8}y = -20$$

$$-\frac{8}{5}\left(-\frac{5}{8}y\right) = -\frac{8}{5}(-20)$$

$$y = 32$$

29.   Simplifying and then solving the equation:

$$-4x - 2x + 3x = 24$$

$$-3x = 24$$

$$-\frac{1}{3}(-3x) = -\frac{1}{3}(24)$$

$$x = -8$$

31.   Simplifying and then solving the equation:

$$4x + 8x - 2x = 15 - 10$$

$$10x = 5$$

$$\frac{1}{10}(10x) = \frac{1}{10}(5)$$

$$x = \frac{1}{2}$$

33.   Simplifying and then solving the equation:

$$-3 - 5 = 3x + 5x - 10x$$

$$-8 = -2x$$

$$-\frac{1}{2}(-8) = -\frac{1}{2}(-2x)$$

$$x = 4$$

35.   Eliminating fractions:

$$18 - 13 = \frac{1}{2}a + \frac{3}{4}a - \frac{5}{8}a$$

$$8(5) = 8\left(\frac{1}{2}a + \frac{3}{4}a - \frac{5}{8}a\right)$$

$$40 = 4a + 6a - 5a$$

$$40 = 5a$$

$$\frac{1}{5}(40) = \frac{1}{5}(5a)$$

$$a = 8$$

37.   Solving by multiplying both sides of the equation by −1:

$$-x = 4$$

$$-1(-x) = -1(4)$$

$$x = -4$$

39. Solving by multiplying both sides of the equation by –1:
$$-x = -4$$
$$-1(-x) = -1(-4)$$
$$x = 4$$

41. Solving by multiplying both sides of the equation by –1:
$$15 = -a$$
$$-1(15) = -1(-a)$$
$$a = -15$$

43. Solving by multiplying both sides of the equation by –1:
$$-y = \frac{1}{2}$$
$$-1(-y) = -1\left(\frac{1}{2}\right)$$
$$y = -\frac{1}{2}$$

45. Solving the equation:
$$3x - 2 = 7$$
$$3x - 2 + 2 = 7 + 2$$
$$3x = 9$$
$$\frac{1}{3}(3x) = \frac{1}{3}(9)$$
$$x = 3$$

47. Solving the equation:
$$2a + 1 = 3$$
$$2a + 1 + (-1) = 3 + (-1)$$
$$2a = 2$$
$$\frac{1}{2}(2a) = \frac{1}{2}(2)$$
$$a = 1$$

49. Eliminating fractions:
$$\frac{1}{8} + \frac{1}{2}x = \frac{1}{4}$$
$$8\left(\frac{1}{8} + \frac{1}{2}x\right) = 8\left(\frac{1}{4}\right)$$
$$1 + 4x = 2$$
$$(-1) + 1 + 4x = (-1) + 2$$
$$4x = 1$$
$$\frac{1}{4}(4x) = \frac{1}{4}(1)$$
$$x = \frac{1}{4}$$

51. Eliminating fractions:
$$6x = 2x - 12$$
$$6x + (-2x) = 2x + (-2x) - 12$$
$$4x = -12$$
$$\frac{1}{4}(4x) = \frac{1}{4}(-12)$$
$$x = -3$$

53. Solving the equation:
$$2y = -4y + 18$$
$$2y + 4y = -4y + 4y + 18$$
$$6y = 18$$
$$\frac{1}{6}(6y) = \frac{1}{6}(18)$$
$$y = 3$$

55. Solving the equation:
$$-7x = -3x - 8$$
$$-7x + 3x = -3x + 3x - 8$$
$$-4x = -8$$
$$-\frac{1}{4}(-4x) = -\frac{1}{4}(-8)$$
$$x = 2$$

57. Solving the equation:

$$8x + 4 = 2x - 5$$
$$8x + (-2x) + 4 = 2x + (-2x) - 5$$
$$6x + 4 = -5$$
$$6x + 4 + (-4) = -5 + (-4)$$
$$6x = -9$$
$$\frac{1}{6}(6x) = \frac{1}{6}(-9)$$
$$x = -\frac{3}{2}$$

59. Solving the equation:

$$x + \frac{1}{2} = \frac{1}{4}x - \frac{5}{8}$$
$$8\left(x + \frac{1}{2}\right) = 8\left(\frac{1}{4}x - \frac{5}{8}\right)$$
$$8x + 4 = 2x - 5$$
$$8x + (-2x) + 4 = 2x + (-2x) - 5$$
$$6x + 4 = -5$$
$$6x + 4 + (-4) = -5 + (-4)$$
$$6x = -9$$
$$\frac{1}{6}(6x) = \frac{1}{6}(-9)$$
$$x = -\frac{3}{2}$$

61. Solving the equation:

$$6m - 3 = m + 2$$
$$6m + (-m) - 3 = m + (-m) + 2$$
$$5m - 3 = 2$$
$$5m - 3 + 3 = 2 + 3$$
$$5m = 5$$
$$\frac{1}{5}(5m) = \frac{1}{5}(5)$$
$$m = 1$$

63. Solving the equation:

$$\frac{1}{2}m - \frac{1}{4} = \frac{1}{12}m + \frac{1}{6}$$
$$12\left(\frac{1}{2}m - \frac{1}{4}\right) = 12\left(\frac{1}{12}m + \frac{1}{6}\right)$$
$$6m - 3 = m + 2$$
$$6m + (-m) - 3 = m + (-m) + 2$$
$$5m - 3 = 2$$
$$5m - 3 + 3 = 2 + 3$$
$$5m = 5$$
$$\frac{1}{5}(5m) = \frac{1}{5}(5)$$
$$m = 1$$

65. Solving the equation:

$$9y + 2 = 6y - 4$$
$$9y + (-6y) + 2 = 6y + (-6y) - 4$$
$$3y + 2 = -4$$
$$3y + 2 + (-2) = -4 + (-2)$$
$$3y = -6$$
$$\frac{1}{3}(3y) = \frac{1}{3}(-6)$$
$$y = -2$$

67. a. Solving the equation:

$$2x = 3$$
$$\frac{1}{2}(2x) = \frac{1}{2}(3)$$
$$x = \frac{3}{2}$$

b. Solving the equation:

$$2 + x = 3$$
$$2 + (-2) + x = 3 + (-2)$$
$$x = 1$$

c. Solving the equation:
$$2x+3=0$$
$$2x+3+(-3)=0+(-3)$$
$$2x=-3$$
$$\frac{1}{2}(2x)=\frac{1}{2}(-3)$$
$$x=-\frac{3}{2}$$

d. Solving the equation:
$$2x+3=-5$$
$$2x+3+(-3)=-5+(-3)$$
$$2x=-8$$
$$\frac{1}{2}(2x)=\frac{1}{2}(-8)$$
$$x=-4$$

e. Solving the equation:
$$2x+3=7x-5$$
$$2x+(-7x)+3=7x+(-7x)-5$$
$$-5x+3=-5$$
$$-5x+3+(-3)=-5+(-3)$$
$$-5x=-8$$
$$-\frac{1}{5}(-5x)=-\frac{1}{5}(-8)$$
$$x=\frac{8}{5}$$

69. Solving the equation:
$$7.5x=1500$$
$$\frac{7.5x}{7.5}=\frac{1500}{7.5}$$
$$x=200$$
The break-even point is 200 tickets.

71. Solving the equation:
$$G-0.21G-0.08G=987.5$$
$$0.71G=987.5$$
$$G\approx1390.85$$
Your gross income is approximately $1,390.85.

73. Solving the equation:
$$2x=4$$
$$\frac{1}{2}(2x)=\frac{1}{2}(4)$$
$$x=2$$

75. Solving the equation:
$$30=5x$$
$$5x=30$$
$$\frac{1}{5}(5x)=\frac{1}{5}(30)$$
$$x=6$$

77. Solving the equation:
$$0.17x=510$$
$$x=\frac{510}{0.17}=3,000$$

79. Simplifying: $3(x-5)+4=3x-15+4=3x-11$

81. Simplifying: $0.09(x+2,000)=0.09x+180$

83. Simplifying: $7-3(2y+1)=7-6y-3=4-6y=-6y+4$

85. Simplifying: $3(2x-5)-(2x-4)=6x-15-2x+4=4x-11$

87. Simplifying: $10x+(-5x)=5x$

89. Simplifying: $0.08x+0.09x=0.17x$

2.4 Solving Linear Equations

1. Solving the equation:
$$2(x+3)=12$$
$$2x+6=12$$
$$2x+6+(-6)=12+(-6)$$
$$2x=6$$
$$\frac{1}{2}(2x)=\frac{1}{2}(6)$$
$$x=3$$

3. Solving the equation:
$$6(x-1)=-18$$
$$6x-6=-18$$
$$6x-6+6=-18+6$$
$$6x=-12$$
$$\frac{1}{6}(6x)=\frac{1}{6}(-12)$$
$$x=-2$$

5. Solving the equation:
$$2(4a+1)=-6$$
$$8a+2=-6$$
$$8a+2+(-2)=-6+(-2)$$
$$8a=-8$$
$$\frac{1}{8}(8a)=\frac{1}{8}(-8)$$
$$a=-1$$

7. Solving the equation:
$$14=2(5x-3)$$
$$14=10x-6$$
$$14+6=10x-6+6$$
$$20=10x$$
$$\frac{1}{10}(20)=\frac{1}{10}(10x)$$
$$x=2$$

9. Solving the equation:
$$-2(3y+5)=14$$
$$-6y-10=14$$
$$-6y-10+10=14+10$$
$$-6y=24$$
$$-\frac{1}{6}(-6y)=-\frac{1}{6}(24)$$
$$y=-4$$

11. Solving the equation:
$$-5(2a+4)=0$$
$$-10a-20=0$$
$$-10a-20+20=0+20$$
$$-10a=20$$
$$-\frac{1}{10}(-10a)=-\frac{1}{10}(20)$$
$$a=-2$$

13. Solving the equation:
$$1=\frac{1}{2}(4x+2)$$
$$1=2x+1$$
$$1+(-1)=2x+1+(-1)$$
$$0=2x$$
$$\frac{1}{2}(0)=\frac{1}{2}(2x)$$
$$x=0$$

15. Solving the equation:
$$3(t-4)+5=-4$$
$$3t-12+5=-4$$
$$3t-7=-4$$
$$3t-7+7=-4+7$$
$$3t=3$$
$$\frac{1}{3}(3t)=\frac{1}{3}(3)$$
$$t=1$$

17. Solving the equation:

$$4(2y+1)-7=1$$
$$8y+4-7=1$$
$$8y-3=1$$
$$8y-3+3=1+3$$
$$8y=4$$
$$\frac{1}{8}(8y)=\frac{1}{8}(4)$$
$$y=\frac{1}{2}$$

19. Solving the equation:

$$\frac{1}{2}(x-3)=\frac{1}{4}(x+1)$$
$$\frac{1}{2}x-\frac{3}{2}=\frac{1}{4}x+\frac{1}{4}$$
$$4\left(\frac{1}{2}x-\frac{3}{2}\right)=4\left(\frac{1}{4}x+\frac{1}{4}\right)$$
$$2x-6=x+1$$
$$2x+(-x)-6=x+(-x)+1$$
$$x-6=1$$
$$x-6+6=1+6$$
$$x=7$$

21. Solving the equation:
$$-0.7(2x-7)=0.3(11-4x)$$
$$-1.4x+4.9=3.3-1.2x$$
$$-1.4x+1.2x+4.9=3.3-1.2x+1.2x$$
$$-0.2x+4.9=3.3$$
$$-0.2x+4.9+(-4.9)=3.3+(-4.9)$$
$$-0.2x=-1.6$$
$$\frac{-0.2x}{-0.2}=\frac{-1.6}{-0.2}$$
$$x=8$$

23. Solving the equation:

$$-2(3y+1)=3(1-6y)-9$$
$$-6y-2=3-18y-9$$
$$-6y-2=-18y-6$$
$$-6y+18y-2=-18y+18y-6$$
$$12y-2=-6$$
$$12y-2+2=-6+2$$
$$12y=-4$$
$$\frac{1}{12}(12y)=\frac{1}{12}(-4)$$
$$y=-\frac{1}{3}$$

25. Solving the equation:
$$\frac{3}{4}(8x-4)+3=\frac{2}{5}(5x+10)-1$$
$$6x-3+3=2x+4-1$$
$$6x=2x+3$$
$$6x+(-2x)=2x+(-2x)+3$$
$$4x=3$$
$$\frac{1}{4}(4x)=\frac{1}{4}(3)$$
$$x=\frac{3}{4}$$

27. Solving the equation:
$$0.06x+0.08(100-x)=6.5$$
$$0.06x+8-0.08x=6.5$$
$$-0.02x+8=6.5$$
$$-0.02x+8+(-8)=6.5+(-8)$$
$$-0.02x=-1.5$$
$$\frac{-0.02x}{-0.02}=\frac{-1.5}{-0.02}$$
$$x=75$$

29. Solving the equation:

$$6 - 5(2a - 3) = 1$$
$$6 - 10a + 15 = 1$$
$$-10a + 21 = 1$$
$$-10a + 21 + (-21) = 1 + (-21)$$
$$-10a = -20$$
$$-\frac{1}{10}(-10a) = -\frac{1}{10}(-20)$$
$$a = 2$$

31. Solving the equation:

$$0.2x - 0.5 = 0.5 - 0.2(2x - 13)$$
$$0.2x - 0.5 = 0.5 - 0.4x + 2.6$$
$$0.2x - 0.5 = -0.4x + 3.1$$
$$0.2x + 0.4x - 0.5 = -0.4x + 0.4x + 3.1$$
$$0.6x - 0.5 = 3.1$$
$$0.6x - 0.5 + 0.5 = 3.1 + 0.5$$
$$0.6x = 3.6$$
$$\frac{0.6x}{0.6} = \frac{3.6}{0.6}$$
$$x = 6$$

33. Solving the equation:

$$2(t - 3) + 3(t - 2) = 28$$
$$2t - 6 + 3t - 6 = 28$$
$$5t - 12 = 28$$
$$5t - 12 + 12 = 28 + 12$$
$$5t = 40$$
$$\frac{1}{5}(5t) = \frac{1}{5}(40)$$
$$t = 8$$

35. Solving the equation:

$$5(x - 2) - (3x + 4) = 3(6x - 8) + 10$$
$$5x - 10 - 3x - 4 = 18x - 24 + 10$$
$$2x - 14 = 18x - 14$$
$$2x + (-18x) - 14 = 18x + (-18x) - 14$$
$$-16x - 14 = -14$$
$$-16x - 14 + 14 = -14 + 14$$
$$-16x = 0$$
$$-\frac{1}{16}(-16x) = -\frac{1}{16}(0)$$
$$x = 0$$

37. Solving the equation:

$$2(5x - 3) - (2x - 4) = 5 - (6x + 1)$$
$$10x - 6 - 2x + 4 = 5 - 6x - 1$$
$$8x - 2 = -6x + 4$$
$$8x + 6x - 2 = -6x + 6x + 4$$
$$14x - 2 = 4$$
$$14x - 2 + 2 = 4 + 2$$
$$14x = 6$$
$$\frac{1}{14}(14x) = \frac{1}{14}(6)$$
$$x = \frac{3}{7}$$

39. Solving the equation:

$$-(3x + 1) - (4x - 7) = 4 - (3x + 2)$$
$$-3x - 1 - 4x + 7 = 4 - 3x - 2$$
$$-7x + 6 = -3x + 2$$
$$-7x + 3x + 6 = -3x + 3x + 2$$
$$-4x + 6 = 2$$
$$-4x + 6 + (-6) = 2 + (-6)$$
$$-4x = -4$$
$$-\frac{1}{4}(-4x) = -\frac{1}{4}(-4)$$
$$x = 1$$

41. Solving the equation:

$$x + (2x - 1) = 2$$
$$3x - 1 = 2$$
$$3x - 1 + 1 = 2 + 1$$
$$3x = 3$$
$$\frac{1}{3}(3x) = \frac{1}{3}(3)$$
$$x = 1$$

43. Solving the equation:

$$x - (3x + 5) = -3$$
$$x - 3x - 5 = -3$$
$$-2x - 5 = -3$$
$$-2x - 5 + 5 = -3 + 5$$
$$-2x = 2$$
$$-\frac{1}{2}(-2x) = -\frac{1}{2}(2)$$
$$x = -1$$

45. Solving the equation:

$$15 = 3(x-1)$$
$$15 = 3x - 3$$
$$15 + 3 = 3x - 3 + 3$$
$$18 = 3x$$
$$\frac{1}{3}(18) = \frac{1}{3}(3x)$$
$$x = 6$$

47. Solving the equation:

$$4x - (-4x + 1) = 5$$
$$4x + 4x - 1 = 5$$
$$8x - 1 = 5$$
$$8x - 1 + 1 = 5 + 1$$
$$8x = 6$$
$$\frac{1}{8}(8x) = \frac{1}{8}(6)$$
$$x = \frac{3}{4}$$

49. Solving the equation:

$$5x - 8(2x - 5) = 7$$
$$5x - 16x + 40 = 7$$
$$-11x + 40 = 7$$
$$-11x + 40 - 40 = 7 - 40$$
$$-11x = -33$$
$$-\frac{1}{11}(-11x) = -\frac{1}{11}(-33)$$
$$x = 3$$

51. Solving the equation:

$$7(2y - 1) - 6y = -1$$
$$14y - 7 - 6y = -1$$
$$8y - 7 = -1$$
$$8y - 7 + 7 = -1 + 7$$
$$8y = 6$$
$$\frac{1}{8}(8y) = \frac{1}{8}(6)$$
$$y = \frac{3}{4}$$

53. Solving the equation:

$$0.2x + 0.5(12 - x) = 3.6$$
$$0.2x + 6 - 0.5x = 3.6$$
$$-0.3x + 6 = 3.6$$
$$-0.3x + 6 - 6 = 3.6 - 6$$
$$-0.3x = -2.4$$
$$x = \frac{-2.4}{-0.3} = 8$$

55. Solving the equation:

$$0.5x + 0.2(18 - x) = 5.4$$
$$0.5x + 3.6 - 0.2x = 5.4$$
$$0.3x + 3.6 = 5.4$$
$$0.3x + 3.6 - 3.6 = 5.4 - 3.6$$
$$0.3x = 1.8$$
$$x = \frac{1.8}{0.3} = 6$$

57. Solving the equation:

$$x + (x + 3)(-3) = x - 3$$
$$x - 3x - 9 = x - 3$$
$$-2x - 9 = x - 3$$
$$-2x - x - 9 = x - x - 3$$
$$-3x - 9 = -3$$
$$-3x - 9 + 9 = -3 + 9$$
$$-3x = 6$$
$$-\frac{1}{3}(-3x) = -\frac{1}{3}(6)$$
$$x = -2$$

59. Solving the equation:

$$5(x + 2) + 3(x - 1) = -9$$
$$5x + 10 + 3x - 3 = -9$$
$$8x + 7 = -9$$
$$8x + 7 - 7 = -9 - 7$$
$$8x = -16$$
$$\frac{1}{8}(8x) = \frac{1}{8}(-16)$$
$$x = -2$$

61. Solving the equation:
$$3(x-3)+2(2x)=5$$
$$3x-9+4x=5$$
$$7x-9=5$$
$$7x-9+9=5+9$$
$$7x=14$$
$$\frac{1}{7}(7x)=\frac{1}{7}(14)$$
$$x=2$$

63. Solving the equation:
$$5(y+2)=4(y+1)$$
$$5y+10=4y+4$$
$$5y-4y+10=4y-4y+4$$
$$y+10=4$$
$$y+10-10=4-10$$
$$y=-6$$

65. Solving the equation:
$$3x+2(x-2)=6$$
$$3x+2x-4=6$$
$$5x-4=6$$
$$5x-4+4=6+4$$
$$5x=10$$
$$\frac{1}{5}(5x)=\frac{1}{5}(10)$$
$$x=2$$

67. Solving the equation:
$$50(x-5)=30(x+5)$$
$$50x-250=30x+150$$
$$50x-30x-250=30x-30x+150$$
$$20x-250=150$$
$$20x-250+250=150+250$$
$$20x=400$$
$$\frac{1}{20}(20x)=\frac{1}{20}(400)$$
$$x=20$$

69. Solving the equation:
$$0.08x+0.09(x+2,000)=860$$
$$0.08x+0.09x+180=860$$
$$0.17x+180=860$$
$$0.17x+180-180=860-180$$
$$0.17x=680$$
$$x=\frac{680}{0.17}=4,000$$

71. Solving the equation:
$$0.10x+0.12(x+500)=214$$
$$0.10x+0.12x+60=214$$
$$0.22x+60=214$$
$$0.22x+60-60=214-60$$
$$0.22x=154$$
$$x=\frac{154}{0.22}=700$$

73. Solving the equation:
$$5x+10(x+8)=245$$
$$5x+10x+80=245$$
$$15x+80=245$$
$$15x+80-80=245-80$$
$$15x=165$$
$$x=\frac{165}{15}=11$$

75. Solving the equation:
$$5x+10(x+3)+25(x+5)=435$$
$$5x+10x+30+25x+125=435$$
$$40x+155=435$$
$$40x+155-155=435-155$$
$$40x=280$$
$$x=\frac{280}{40}=7$$

77. a. Solving the equation:
$$4x-5=0$$
$$4x-5+5=0+5$$
$$4x=5$$
$$\frac{1}{4}(4x)=\frac{1}{4}(5)$$
$$x=\frac{5}{4}=1.25$$

b. Solving the equation:
$$4x-5=25$$
$$4x-5+5=25+5$$
$$4x=30$$
$$\frac{1}{4}(4x)=\frac{1}{4}(30)$$
$$x=\frac{15}{2}=7.5$$

c.   Adding: $(4x-5)+(2x+25)=6x+20$

d.   Solving the equation:
$$4x-5=2x+25$$
$$4x-2x-5=2x-2x+25$$
$$2x-5=25$$
$$2x-5+5=25+5$$
$$2x=30$$
$$\frac{1}{2}(2x)=\frac{1}{2}(30)$$
$$x=15$$

e.   Multiplying: $4(x-5)=4x-20$

f.   Solving the equation:
$$4(x-5)=2x+25$$
$$4x-20=2x+25$$
$$4x-2x-20=2x-2x+25$$
$$2x-20=25$$
$$2x-20+20=25+20$$
$$2x=45$$
$$\frac{1}{2}(2x)=\frac{1}{2}(45)$$
$$x=\frac{45}{2}=22.5$$

79.   Solving the equation:
$$40=2x+12$$
$$2x+12=40$$
$$2x=28$$
$$x=14$$

81.   Solving the equation:
$$12+2y=6$$
$$2y=-6$$
$$y=-3$$

83.   Solving the equation:
$$24x=6$$
$$x=\frac{6}{24}=\frac{1}{4}$$

85.   Solving the equation:
$$70=x\bullet 210$$
$$x=\frac{70}{210}=\frac{1}{3}$$

87.   Simplifying: $\frac{1}{2}(-3x+6)=\frac{1}{2}(-3x)+\frac{1}{2}(6)=-\frac{3}{2}x+3$

2.5 Formulas

1. Using the perimeter formula:
$$P = 2l + 2w$$
$$300 = 2l + 2(50)$$
$$300 = 2l + 100$$
$$200 = 2l$$
$$l = 100$$
The length is 100 feet.

3. Substituting $x = 3$:
$$2(3) + 3y = 6$$
$$6 + 3y = 6$$
$$6 + (-6) + 3y = 6 + (-6)$$
$$3y = 0$$
$$y = 0$$

5. Substituting $x = 0$:
$$2(0) + 3y = 6$$
$$0 + 3y = 6$$
$$3y = 6$$
$$y = 2$$

7. Substituting $y = 2$:
$$2x - 5(2) = 20$$
$$2x - 10 = 20$$
$$2x - 10 + 10 = 20 + 10$$
$$2x = 30$$
$$x = 15$$

9. Substituting $y = 0$:
$$2x - 5(0) = 20$$
$$2x - 0 = 20$$
$$2x = 20$$
$$x = 10$$

11. Substituting $x = -2$: $y = (-2 + 1)^2 - 3 = (-1)^2 - 3 = 1 - 3 = -2$

13. Substituting $x = 1$: $y = (1 + 1)^2 - 3 = (2)^2 - 3 = 4 - 3 = 1$

15. **a.** Substituting $x = 10$: $y = \dfrac{20}{10} = 2$ **b.** Substituting $x = 5$: $y = \dfrac{20}{5} = 4$

17. **a.** Substituting $y = 15$ and $x = 3$:
$$15 = K(3)$$
$$K = \frac{15}{3} = 5$$
b. Substituting $y = 72$ and $x = 4$:
$$72 = K(4)$$
$$K = \frac{72}{4} = 18$$

19. Solving for l:
$$lw = A$$
$$\frac{lw}{w} = \frac{A}{w}$$
$$l = \frac{A}{w}$$

21. Solving for h:
$$lwh = V$$
$$\frac{lwh}{lw} = \frac{V}{lw}$$
$$h = \frac{V}{lw}$$

23. Solving for a:
$$a + b + c = P$$
$$a + b + c - b - c = P - b - c$$
$$a = P - b - c$$

25. Solving for x:
$$x - 3y = -1$$
$$x - 3y + 3y = -1 + 3y$$
$$x = 3y - 1$$

27. Solving for y:

$$-3x + y = 6$$
$$-3x + 3x + y = 6 + 3x$$
$$y = 3x + 6$$

29. Solving for y:

$$2x + 3y = 6$$
$$-2x + 2x + 3y = -2x + 6$$
$$3y = -2x + 6$$
$$\frac{1}{3}(3y) = \frac{1}{3}(-2x + 6)$$
$$y = -\frac{2}{3}x + 2$$

31. Solving for y:

$$y - 3 = -2(x + 4)$$
$$y - 3 = -2x - 8$$
$$y = -2x - 5$$

33. Solving for y:

$$y - 3 = -\frac{2}{3}(x + 3)$$
$$y - 3 = -\frac{2}{3}x - 2$$
$$y = -\frac{2}{3}x + 1$$

35. Solving for w:

$$2l + 2w = P$$
$$2l - 2l + 2w = P - 2l$$
$$2w = P - 2l$$
$$\frac{2w}{2} = \frac{P - 2l}{2}$$
$$w = \frac{P - 2l}{2}$$

37. Solving for v:

$$vt + 16t^2 = h$$
$$vt + 16t^2 - 16t^2 = h - 16t^2$$
$$vt = h - 16t^2$$
$$\frac{vt}{t} = \frac{h - 16t^2}{t}$$
$$v = \frac{h - 16t^2}{t}$$

39. Solving for h:

$$\pi r^2 + 2\pi rh = A$$
$$\pi r^2 - \pi r^2 + 2\pi rh = A - \pi r^2$$
$$2\pi rh = A - \pi r^2$$
$$\frac{2\pi rh}{2\pi r} = \frac{A - \pi r^2}{2\pi r}$$
$$h = \frac{A - \pi r^2}{2\pi r}$$

41. a. Solving for y:

$$\frac{y - 1}{x} = \frac{3}{5}$$
$$5y - 5 = 3x$$
$$5y = 3x + 5$$
$$y = \frac{3}{5}x + 1$$

b. Solving for y:

$$\frac{y - 2}{x} = \frac{1}{2}$$
$$2y - 4 = x$$
$$2y = x + 4$$
$$y = \frac{1}{2}x + 2$$

c. Solving for y:

$$\frac{y - 3}{x} = 4$$
$$y - 3 = 4x$$
$$y = 4x + 3$$

43. Solving for y:

$$\frac{x}{7} - \frac{y}{3} = 1$$

$$-\frac{x}{7} + \frac{x}{7} - \frac{y}{3} = -\frac{x}{7} + 1$$

$$-\frac{y}{3} = -\frac{x}{7} + 1$$

$$-3\left(-\frac{y}{3}\right) = -3\left(-\frac{x}{7} + 1\right)$$

$$y = \frac{3}{7}x - 3$$

45. Solving for y:

$$-\frac{1}{4}x + \frac{1}{8}y = 1$$

$$-\frac{1}{4}x + \frac{1}{4}x + \frac{1}{8}y = 1 + \frac{1}{4}x$$

$$\frac{1}{8}y = \frac{1}{4}x + 1$$

$$8\left(\frac{1}{8}y\right) = 8\left(\frac{1}{4}x + 1\right)$$

$$y = 2x + 8$$

47. The complement of $30°$ is $90° - 30° = 60°$, and the supplement is $180° - 30° = 150°$.

49. The complement of $45°$ is $90° - 45° = 45°$, and the supplement is $180° - 45° = 135°$.

51. Translating into an equation and solving:

$$x = 0.25 \cdot 40$$
$$x = 10$$

The number 10 is 25% of 40.

53. Translating into an equation and solving:

$$x = 0.12 \cdot 2000$$
$$x = 240$$

The number 240 is 12% of 2000.

55. Translating into an equation and solving:

$$x \cdot 28 = 7$$
$$28x = 7$$
$$\frac{1}{28}(28x) = \frac{1}{28}(7)$$
$$x = 0.25 = 25\%$$

The number 7 is 25% of 28.

57. Translating into an equation and solving:

$$x \cdot 40 = 14$$
$$40x = 14$$
$$\frac{1}{40}(40x) = \frac{1}{40}(14)$$
$$x = 0.35 = 35\%$$

The number 14 is 35% of 40.

59. Translating into an equation and solving:

$$0.50 \cdot x = 32$$
$$\frac{0.50x}{0.50} = \frac{32}{0.50}$$
$$x = 64$$

The number 32 is 50% of 64.

61. Translating into an equation and solving:

$$0.12 \cdot x = 240$$
$$\frac{0.12x}{0.12} = \frac{240}{0.12}$$
$$x = 2,000$$

The number 240 is 12% of 2,000.

63. Substituting $F = 212$: $C = \frac{5}{9}(212 - 32) = \frac{5}{9}(180) = 100°C$.

This value agrees with the information in Table 1.

65. Substituting $F = 68$: $C = \frac{5}{9}(68 - 32) = \frac{5}{9}(36) = 20°C$.

This value agrees with the information in Table 1.

67. Solving for C:

$$\frac{9}{5}C + 32 = F$$

$$\frac{9}{5}C + 32 - 32 = F - 32$$

$$\frac{9}{5}C = F - 32$$

$$\frac{5}{9}\left(\frac{9}{5}C\right) = \frac{5}{9}(F - 32)$$

$$C = \frac{5}{9}(F - 32)$$

69. Budd's estimate would be: $F = 2(30) + 30 = 60 + 30 = 90°\text{F}$

The actual conversion would be: $F = \frac{9}{5}(30) + 32 = 54 + 32 = 86°\text{F}$

Budd's estimate is 4°F too high.

71. Solving for r:

$$2\pi r = C$$

$$2 \bullet \frac{22}{7}r = 44$$

$$\frac{44}{7}r = 44$$

$$\frac{7}{44}\left(\frac{44}{7}r\right) = \frac{7}{44}(44)$$

$$r = 7 \text{ meters}$$

73. Solving for r:

$$2\pi r = C$$

$$2 \bullet 3.14r = 9.42$$

$$6.28r = 9.42$$

$$\frac{6.28r}{6.28} = \frac{9.42}{6.28}$$

$$r = 1.5 \text{ inches}$$

75. Solving for h:

$$\pi r^2 h = V$$

$$\frac{22}{7}\left(\frac{7}{22}\right)^2 h = 42$$

$$\frac{7}{22}h = 42$$

$$\frac{22}{7}\left(\frac{7}{22}h\right) = \frac{22}{7}(42)$$

$$h = 132 \text{ feet}$$

77. Solving for h:

$$\pi r^2 h = V$$

$$3.14(3)^2 h = 6.28$$

$$28.26h = 6.28$$

$$\frac{28.26h}{28.26} = \frac{6.28}{28.26}$$

$$h = \frac{2}{9} \text{ centimeters}$$

79. We need to find what percent of 150 is 90:

$$x \bullet 150 = 90$$

$$\frac{1}{150}(150x) = \frac{1}{150}(90)$$

$$x = 0.60 = 60\%$$

So 60% of the calories in one serving of vanilla ice cream are fat calories.

81. We need to find what percent of 98 is 26:
$$x \cdot 98 = 26$$
$$\frac{1}{98}(98x) = \frac{1}{98}(26)$$
$$x \approx 0.265 = 26.5\%$$
So 26.5% of one serving of frozen yogurt are carbohydrates.

83. The sum of 4 and 1 is 5.   85. The difference of 6 and 2 is 4.

87. The difference of a number and 5 is –12.

89. The sum of a number and 3 is four times the difference of that number and 3.

91. An equivalent expression is: $2(6+3) = 2(9) = 18$

93. An equivalent expression is: $2(5)+3 = 10+3 = 13$

95. An equivalent expression is: $x+5 = 13$ 97. An equivalent expression is: $5(x+7) = 30$

2.6 Applications

1. Let x represent the number. The equation is:
$$x+5 = 13$$
$$x = 8$$
The number is 8.

3. Let x represent the number. The equation is:
$$2x+4 = 14$$
$$2x = 10$$
$$x = 5$$
The number is 5.

5. Let x represent the number. The equation is:
$$5(x+7) = 30$$
$$5x+35 = 30$$
$$5x = -5$$
$$x = -1$$
The number is –1.

7. Let x and $x+2$ represent the two numbers. The equation is:
$$x+x+2 = 8$$
$$2x+2 = 8$$
$$2x = 6$$
$$x = 3$$
$$x+2 = 5$$
The two numbers are 3 and 5.

9. Let x and $3x-4$ represent the two numbers. The equation is:
$$(x+3x-4)+5 = 25$$
$$4x+1 = 25$$
$$4x = 24$$
$$x = 6$$
$$3x-4 = 3(6)-4 = 14$$
The two numbers are 6 and 14.

11. Completing the table:

	Four Years Ago	Now
Shelly	$x+3-4=x-1$	$x+3$
Michele	$x-4$	x

The equation is:
$$x-1+x-4=67$$
$$2x-5=67$$
$$2x=72$$
$$x=36$$
$$x+3=39$$
Shelly is 39 and Michele is 36.

13. Completing the table:

	Three Years Ago	Now
Cody	$2x-3$	$2x$
Evan	$x-3$	x

The equation is:
$$2x-3+x-3=27$$
$$3x-6=27$$
$$3x=33$$
$$x=11$$
$$2x=22$$
Evan is 11 and Cody is 22.

15. Completing the table:

	Five Years Ago	Now
Fred	$x+4-5=x-1$	$x+4$
Barney	$x-5$	x

The equation is:
$$x-1+x-5=48$$
$$2x-6=48$$
$$2x=54$$
$$x=27$$
$$x+4=31$$
Barney is 27 and Fred is 31.

17. Completing the table:

	Now	Three Years from Now
Jack	$2x$	$2x+3$
Lacy	x	$x+3$

The equation is:
$$2x+3+x+3=54$$
$$3x+6=54$$
$$3x=48$$
$$x=16$$
$$2x=32$$
Lacy is 16 and Jack is 32.

19. Completing the table:

	Now	Two Years from Now
Pat	$x+20$	$x+20+2=x+22$
Patrick	x	$x+2$

The equation is:
$$x+22=2(x+2)$$
$$x+22=2x+4$$
$$22=x+4$$
$$x=18$$
$$x+20=38$$
Patrick is 18 and Pat is 38.

21. Using the formula $P=4s$, the equation is:
$$4s=36$$
$$s=9$$
The length of each side is 9 inches.

23. Using the formula $P=4s$, the equation is:
$$4s=60$$
$$s=15$$
The length of each side is 15 feet.

25.   Let x, $3x$, and $x + 7$ represent the sides of the triangle. The equation is:
$$x + 3x + x + 7 = 62$$
$$5x + 7 = 62$$
$$5x = 55$$
$$x = 11$$
$$x + 7 = 18$$
$$3x = 33$$
The sides are 11 feet, 18 feet, and 33 feet.

27.   Let x, $2x$, and $2x - 12$ represent the sides of the triangle. The equation is:
$$x + 2x + 2x - 12 = 53$$
$$5x - 12 = 53$$
$$5x = 65$$
$$x = 13$$
$$2x = 26$$
$$2x - 12 = 14$$
The sides are 13 feet, 14 feet, and 26 feet.

29.   Let w represent the width and $w + 5$ represent the length. The equation is:
$$2w + 2(w + 5) = 34$$
$$2w + 2w + 10 = 34$$
$$4w + 10 = 34$$
$$4w = 24$$
$$w = 6$$
$$w + 5 = 11$$
The length is 11 inches and the width is 6 inches.

31.   Let w represent the width and $2w + 7$ represent the length. The equation is:
$$2w + 2(2w + 7) = 68$$
$$2w + 4w + 14 = 68$$
$$6w + 14 = 68$$
$$6w = 54$$
$$w = 9$$
$$2w + 7 = 2(9) + 7 = 25$$
The length is 25 inches and the width is 9 inches.

33.   Let w represent the width and $3w + 6$ represent the length. The equation is:
$$2w + 2(3w + 6) = 36$$
$$2w + 6w + 12 = 36$$
$$8w + 12 = 36$$
$$8w = 24$$
$$w = 3$$
$$3w + 6 = 3(3) + 6 = 15$$
The length is 15 feet and the width is 3 feet.

35.   Completing the table:

	Dimes	Quarters
Number	x	$x + 5$
Value (cents)	$10(x)$	$25(x + 5)$

37.   Completing the table:

	Nickels	Quarters
Number	$x + 15$	x
Value (cents)	$5(x + 15)$	$25(x)$

The equation is:

$$10(x)+25(x+5)=440$$
$$10x+25x+125=440$$
$$35x+125=440$$
$$35x=315$$
$$x=9$$
$$x+5=14$$

Marissa has 9 dimes and 14 quarters.

39. Completing the table:

	Nickels	Dimes
Number	x	$x+9$
Value (cents)	$5(x)$	$10(x+9)$

The equation is:

$$5(x)+10(x+9)=210$$
$$5x+10x+90=210$$
$$15x+90=210$$
$$15x=120$$
$$x=8$$
$$x+9=17$$

Sue has 8 nickels and 17 dimes.

43. Completing the table:

	Nickels	Dimes	Quarters
Number	x	$x+6$	$2x$
Value (cents)	$5(x)$	$10(x+6)$	$25(2x)$

The equation is:

$$5(x)+10(x+6)+25(2x)=255$$
$$5x+10x+60+50x=255$$
$$65x+60=255$$
$$65x=195$$
$$x=3$$
$$x+6=9$$
$$2x=6$$

Cory has 3 nickels, 9 dimes, and 6 quarters.

45. Simplifying: $x+2x+2x=5x$

49. Simplifying: $0.09(x+2,000)=0.09x+180$

51. Solving the equation:

$$0.05x+0.06(x-1,500)=570$$
$$0.05x+0.06x-90=570$$
$$0.11x=660$$
$$x=6,000$$

The equation is:

$$5(x+15)+25(x)=435$$
$$5x+75+25x=435$$
$$30x+75=435$$
$$30x=360$$
$$x=12$$
$$x+15=27$$

Tanner has 12 quarters and 27 nickels.

41. Completing the table:

	Nickels	Dimes	Quarters
Number	x	$x+3$	$x+5$
Value (cents)	$5(x)$	$10(x+3)$	$25(x+5)$

The equation is:

$$5(x)+10(x+3)+25(x+5)=435$$
$$5x+10x+30+25x+125=435$$
$$40x+155=435$$
$$40x=280$$
$$x=7$$
$$x+3=10$$
$$x+5=12$$

Katie has 7 nickels, 10 dimes, and 12 quarters.

47. Simplifying: $x+0.075x=1.075x$

53. Solving the equation:

$$x+2x+3x=180$$
$$6x=180$$
$$x=30$$

2.7 More Applications

1. Let x and $x + 1$ represent the two numbers. The equation is:
$$x + x + 1 = 11$$
$$2x + 1 = 11$$
$$2x = 10$$
$$x = 5$$
$$x + 1 = 6$$
The numbers are 5 and 6.

3. Let x and $x + 1$ represent the two numbers. The equation is:
$$x + x + 1 = -9$$
$$2x + 1 = -9$$
$$2x = -10$$
$$x = -5$$
$$x + 1 = -4$$
The numbers are -5 and -4.

5. Let x and $x + 2$ represent the two numbers. The equation is:
$$x + x + 2 = 28$$
$$2x + 2 = 28$$
$$2x = 26$$
$$x = 13$$
$$x + 2 = 15$$
The numbers are 13 and 15.

7. Let x and $x + 2$ represent the two numbers. The equation is:
$$x + x + 2 = 106$$
$$2x + 2 = 106$$
$$2x = 104$$
$$x = 52$$
$$x + 2 = 54$$
The numbers are 52 and 54.

9. Let x and $x + 2$ represent the two numbers. The equation is:
$$x + x + 2 = -30$$
$$2x + 2 = -30$$
$$2x = -322$$
$$x = -16$$
$$x + 2 = -14$$
The numbers are -16 and -14.

11. Let x, $x + 2$, and $x + 4$ represent the three numbers. The equation is:
$$x + x + 2 + x + 4 = 57$$
$$3x + 6 = 57$$
$$3x = 51$$
$$x = 17$$
$$x + 2 = 19$$
$$x + 4 = 21$$
The numbers are 17, 19, and 21.

13. Let x, $x + 2$, and $x + 4$ represent the three numbers. The equation is:

$$x + x + 2 + x + 4 = 132$$
$$3x + 6 = 132$$
$$3x = 126$$
$$x = 42$$
$$x + 2 = 44$$
$$x + 4 = 46$$

The numbers are 42, 44, and 46.

15. Completing the table:

	Dollars Invested at 8%	Dollars Invested at 9%
Number of	x	$x + 2000$
Interest on	$0.08(x)$	$0.09(x + 2000)$

The equation is:

$$0.08(x) + 0.09(x + 2000) = 860$$
$$0.08x + 0.09x + 180 = 860$$
$$0.17x + 180 = 860$$
$$0.17x = 680$$
$$x = 4,000$$
$$x + 2,000 = 6,000$$

You have \$4,000 invested at 8% and \$6,000 invested at 9%.

17. Completing the table:

	Dollars Invested at 10%	Dollars Invested at 12%
Number of	x	$x + 500$
Interest on	$0.10(x)$	$0.12(x + 500)$

The equation is:

$$0.10(x) + 0.12(x + 500) = 214$$
$$0.10x + 0.12x + 60 = 214$$
$$0.22x + 60 = 214$$
$$0.22x = 154$$
$$x = 700$$
$$x + 500 = 1,200$$

Tyler has \$700 invested at 10% and \$1,200 invested at 12%.

19. Completing the table:

	Dollars Invested at 8%	Dollars Invested at 9%	Dollars Invested at 10%
Number of	x	$2x$	$3x$
Interest on	$0.08(x)$	$0.09(2x)$	$0.10(3x)$

The equation is:

$$0.08(x) + 0.09(2x) + 0.10(3x) = 280$$
$$0.08x + 0.18x + 0.30x = 280$$
$$0.56x = 280$$
$$x = 500$$
$$2x = 1,000$$
$$3x = 1,500$$

She has \$500 invested at 8%, \$1,000 invested at 9%, and \$1,500 invested at 10%.

21.   Let x represent the measure of the two equal angles, so $x + x = 2x$ represents the measure of the third angle. Since the sum of the three angles is 180°, the equation is:

$$x + x + 2x = 180°$$
$$4x = 180°$$
$$x = 45°$$
$$2x = 90°$$

The measures of the three angles are 45°, 45°, and 90°.

23.   Let x represent the measure of the largest angle. Then $\frac{1}{5}x$ represents the measure of the smallest angle, and $2\left(\frac{1}{5}x\right) = \frac{2}{5}x$ represents the measure of the other angle. Since the sum of the three angles is 180°, the equation is:

$$x + \frac{1}{5}x + \frac{2}{5}x = 180°$$
$$\frac{5}{5}x + \frac{1}{5}x + \frac{2}{5}x = 180°$$
$$\frac{8}{5}x = 180°$$
$$x = 112.5°$$
$$\frac{1}{5}x = 22.5°$$
$$\frac{2}{5}x = 45°$$

The measures of the three angles are 22.5°, 45°, and 112.5°.

25.   Let x represent the measure of the other acute angle, and 90° is the measure of the right angle. Since the sum of the three angles is 180°, the equation is:

$$x + 37° + 90° = 180°$$
$$x + 127° = 180°$$
$$x = 53°$$

The other two angles are 53° and 90°.

27.   Let x represent the measure of the smallest angle, so $x + 20$ represents the measure of the second angle and $2x$ represents the measure of the third angle. Since the sum of the three angles is 180°, the equation is:

$$x + x + 20 + 2x = 180°$$
$$4x + 20 = 180°$$
$$4x = 160°$$
$$x = 40°$$
$$x + 20 = 60°$$
$$2x = 80°$$

The measures of the three angles are 40°, 60°, and 80°.

29. Completing the table:

	Adult	Child
Number	x	$x+6$
Income	$6(x)$	$4(x+6)$

The equation is:
$$6(x)+4(x+6)=184$$
$$6x+4x+24=184$$
$$10x+24=184$$
$$10x=160$$
$$x=16$$
$$x+6=22$$
Miguel sold 16 adult and 22 children's tickets.

31. Let x represent the total minutes for the call. Then \$0.41 is charged for the first minute, and \$0.32 is charged for the additional $x-1$ minutes. The equation is:
$$0.41(1)+0.32(x-1)=5.21$$
$$0.41+0.32x-0.32=5.21$$
$$0.32x+0.09=5.21$$
$$0.32x=5.12$$
$$x=16$$
The call was 16 minutes long.

33. Let x represent the hours Jo Ann worked that week. Then \$12/hour is paid for the first 35 hours and \$18/hour is paid for the additional $x-35$ hours. The equation is:
$$12(35)+18(x-35)=492$$
$$420+18x-630=492$$
$$18x-210=492$$
$$18x=702$$
$$x=39$$
Jo Ann worked 39 hours that week.

35. Let x and $x+2$ represent the two office numbers. The equation is:
$$x+x+2=14,660$$
$$2x+2=14,660$$
$$2x=14,658$$
$$x=7329$$
$$x+2=7331$$
They are in offices 7329 and 7331.

37. Let x represent Kendra's age and $x+2$ represent Marissa's age. The equation is:
$$x+2+2x=26$$
$$3x+2=26$$
$$3x=24$$
$$x=8$$
$$x+2=10$$
Kendra is 8 years old and Marissa is 10 years old.

39. For Jeff, the total time traveled is $\dfrac{425 \text{ miles}}{55 \text{ miles/hour}} \approx 7.72 \text{ hours} \approx 463 \text{ minutes}$. Since he left at 11:00 AM, he will arrive at 6:43 PM. For Carla, the total time traveled is $\dfrac{425 \text{ miles}}{65 \text{ miles/hour}} \approx 6.54 \text{ hours} \approx 392 \text{ minutes}$. Since she left at 1:00 PM, she will arrive at 7:32 PM. Thus Jeff will arrive in Lake Tahoe first.

41. Since $\dfrac{1}{5}$ mile = 0.2 mile, the taxi charge is \$1.25 for the first $\dfrac{1}{5}$ mile and \$0.25 per fifth mile for the remaining 7.3 miles. Since $7.3 \text{ miles} = \dfrac{7.3}{0.2} = 36.5 \text{ fifths}$, the total charge is:
$$\$1.25 + \$0.25(36.5) \approx \$10.38$$

43. Let w represent the width and $w + 2$ represent the length. The equation is:
$$2w + 2(w + 2) = 44$$
$$2w + 2w + 4 = 44$$
$$4w + 4 = 44$$
$$4w = 40$$
$$w = 10$$
$$w + 2 = 12$$
The length is 12 meters and the width is 10 meters.

45. Let x, $x + 1$, and $x + 2$ represent the measures of the three angles. Since the sum of the three angles is 180°, the equation is:
$$x + x + 1 + x + 2 = 180°$$
$$3x + 3 = 180°$$
$$3x = 177°$$
$$x = 59°$$
$$x + 1 = 60°$$
$$x + 2 = 61°$$
The measures of the three angles are 59°, 60°, and 61°.

47. If all 36 people are Elk's Lodge members (which would be the least amount), the cost of the lessons would be $\$3(36) = \$108$. Since half of the money is paid to Ike and Nancy, the least amount they could make is $\dfrac{1}{2}(\$108) = \$54$.

49. Yes. The total receipts were \$160, which is possible if there were 10 Elk's members and 26 nonmembers. Computing the total receipts: $10(\$3) + 26(\$5) = \$30 + \$130 = \$160$

51. a. Solving the equation:
$$x - 3 = 6$$
$$x = 9$$

b. Solving the equation:
$$x + 3 = 6$$
$$x = 3$$

c. Solving the equation:
$$-x - 3 = 6$$
$$-x = 9$$
$$x = -9$$

d. Solving the equation:
$$-x + 3 = 6$$
$$-x = 3$$
$$x = -3$$

53. **a.** Solving the equation:
$$\frac{x}{4} = -2$$
$$x = -2(4) = -8$$

b. Solving the equation:
$$-\frac{x}{4} = -2$$
$$x = -2(-4) = 8$$

c. Solving the equation:
$$\frac{x}{4} = 2$$
$$x = 2(4) = 8$$

d. Solving the equation:
$$-\frac{x}{4} = 2$$
$$x = 2(-4) = -8$$

55. Solving the equation:
$$2.5x - 3.48 = 4.9x + 2.07$$
$$-2.4x - 3.48 = 2.07$$
$$-2.4x = 5.55$$
$$x = -2.3125$$

56. Solving the equation:
$$2(1 - 3x) + 4 = 4x - 14$$
$$2 - 6x + 4 = 4x - 14$$
$$6 - 10x = -14$$
$$-10x = -20$$
$$x = 2$$

57. Solving the equation:
$$3(x - 4) = -2$$
$$3x - 12 = -2$$
$$3x = 10$$
$$x = \frac{10}{3}$$

2.8 Linear Inequalities

1. Solving the inequality:
$$x - 5 < 7$$
$$x - 5 + 5 < 7 + 5$$
$$x < 12$$
Graphing the solution set:

3. Solving the inequality:
$$a - 4 \le 8$$
$$a - 4 + 4 \le 8 + 4$$
$$a \le 12$$
Graphing the solution set:

5. Solving the inequality:
$$x - 4.3 > 8.7$$
$$x - 4.3 + 4.3 > 8.7 + 4.3$$
$$x > 13$$
Graphing the solution set:

7. Solving the inequality:
$$y + 6 \ge 10$$
$$y + 6 + (-6) \ge 10 + (-6)$$
$$y \ge 4$$
Graphing the solution set:

9. Solving the inequality:
$$2 < x - 7$$
$$2 + 7 < x - 7 + 7$$
$$9 < x$$
$$x > 9$$
Graphing the solution set:

11. Solving the inequality:
$$3x < 6$$
$$\frac{1}{3}(3x) < \frac{1}{3}(6)$$
$$x < 2$$
Graphing the solution set:

13. Solving the inequality:

$$5a \le 25$$
$$\frac{1}{5}(5a) \le \frac{1}{5}(25)$$
$$a \le 5$$

Graphing the solution set:

```
◄─────────────────●──────────►
            0      5
```

15. Solving the inequality:

$$\frac{x}{3} > 5$$
$$3\left(\frac{x}{3}\right) > 3(5)$$
$$x > 15$$

Graphing the solution set:

```
◄──────────┼──────────○──────►
           0          15
```

17. Solving the inequality:

$$-2x > 6$$
$$-\frac{1}{2}(-2x) < -\frac{1}{2}(6)$$
$$x < -3$$

Graphing the solution set:

```
◄──────○───────┼──────────────►
      -3        0
```

19. Solving the inequality:

$$-3x \ge -18$$
$$-\frac{1}{3}(-3x) \le -\frac{1}{3}(-18)$$
$$x \le 6$$

Graphing the solution set:

```
◄──────────────┼──────────●───►
               0          6
```

21. Solving the inequality:

$$-\frac{x}{5} \le 10$$
$$-5\left(-\frac{x}{5}\right) \ge -5(10)$$
$$x \ge -50$$

Graphing the solution set:

```
◄──────●────────┼─────────────►
     -50         0
```

23. Solving the inequality:

$$-\frac{2}{3}y > 4$$
$$-\frac{3}{2}\left(-\frac{2}{3}y\right) < -\frac{3}{2}(4)$$
$$y < -6$$

Graphing the solution set:

```
◄──────○────────┼─────────────►
      -6         0
```

25. Solving the inequality:

$$2x - 3 < 9$$
$$2x - 3 + 3 < 9 + 3$$
$$2x < 12$$
$$\frac{1}{2}(2x) < \frac{1}{2}(12)$$
$$x < 6$$

Graphing the solution set:

```
◄──────────┼──────────○──────►
           0          6
```

27. Solving the inequality:

$$-\frac{1}{5}y - \frac{1}{3} \le \frac{2}{3}$$
$$-\frac{1}{5}y - \frac{1}{3} + \frac{1}{3} \le \frac{2}{3} + \frac{1}{3}$$
$$-\frac{1}{5}y \le 1$$
$$-5\left(-\frac{1}{5}y\right) \ge -5(1)$$
$$y \ge -5$$

Graphing the solution set:

```
◄──────●────────┼─────────────►
      -5         0
```

29. Solving the inequality:

$$-7.2x + 1.8 > -19.8$$
$$-7.2x + 1.8 - 1.8 > -19.8 - 1.8$$
$$-7.2x > -21.6$$
$$\frac{-7.2x}{-7.2} < \frac{-21.6}{-7.2}$$
$$x < 3$$

Graphing the solution set:

31. Solving the inequality:

$$\frac{2}{3}x - 5 \le 7$$
$$\frac{2}{3}x - 5 + 5 \le 7 + 5$$
$$\frac{2}{3}x \le 12$$
$$\frac{3}{2}\left(\frac{2}{3}x\right) \le \frac{3}{2}(12)$$
$$x \le 18$$

Graphing the solution set:

33. Solving the inequality:

$$-\frac{2}{5}a - 3 > 5$$
$$-\frac{2}{5}a - 3 + 3 > 5 + 3$$
$$-\frac{2}{5}a > 8$$
$$-\frac{5}{2}\left(-\frac{2}{5}a\right) < -\frac{5}{2}(8)$$
$$a < -20$$

Graphing the solution set:

35. Solving the inequality:

$$5 - \frac{3}{5}y > -10$$
$$-5 + 5 - \frac{3}{5}y > -5 + (-10)$$
$$-\frac{3}{5}y > -15$$
$$-\frac{5}{3}\left(-\frac{3}{5}y\right) < -\frac{5}{3}(-15)$$
$$y < 25$$

Graphing the solution set:

37. Solving the inequality:

$$0.3(a + 1) \le 1.2$$
$$0.3a + 0.3 \le 1.2$$
$$0.3a + 0.3 + (-0.3) \le 1.2 + (-0.3)$$
$$0.3a \le 0.9$$
$$\frac{0.3a}{0.3} \le \frac{0.9}{0.3}$$
$$a \le 3$$

Graphing the solution set:

39. Solving the inequality:

$$2(5 - 2x) \le -20$$
$$10 - 4x \le -20$$
$$-10 + 10 - 4x \le -10 + (-20)$$
$$-4x \le -30$$
$$-\frac{1}{4}(-4x) \ge -\frac{1}{4}(-30)$$
$$x \ge \frac{15}{2}$$

Graphing the solution set:

41. Solving the inequality:

$$3x - 5 > 8x$$
$$-3x + 3x - 5 > -3x + 8x$$
$$-5 > 5x$$
$$\frac{1}{5}(-5) > \frac{1}{5}(5x)$$
$$-1 > x$$
$$x < -1$$

Graphing the solution set:

43. First multiply by 6 to clear the inequality of fractions:

$$\frac{1}{3}y - \frac{1}{2} \le \frac{5}{6}y + \frac{1}{2}$$

$$6\left(\frac{1}{3}y - \frac{1}{2}\right) \le 6\left(\frac{5}{6}y + \frac{1}{2}\right)$$

$$2y - 3 \le 5y + 3$$

$$-5y + 2y - 3 \le -5y + 5y + 3$$

$$-3y - 3 \le 3$$

$$-3y - 3 + 3 \le 3 + 3$$

$$-3y \le 6$$

$$-\frac{1}{3}(-3y) \ge -\frac{1}{3}(6)$$

$$y \ge -2$$

Graphing the solution set:

45. Solving the inequality:

$$-2.8x + 8.4 < -14x - 2.8$$

$$-2.8x + 14x + 8.4 < -14x + 14x - 2.8$$

$$11.2x + 8.4 < -2.8$$

$$11.2x + 8.4 - 8.4 < -2.8 - 8.4$$

$$11.2x < -11.2$$

$$\frac{11.2x}{11.2} < \frac{-11.2}{11.2}$$

$$x < -1$$

Graphing the solution set:

47. Solving the inequality:

$$3(m-2) - 4 \ge 7m + 14$$

$$3m - 6 - 4 \ge 7m + 14$$

$$3m - 10 \ge 7m + 14$$

$$-7m + 3m - 10 \ge -7m + 7m + 14$$

$$-4m - 10 \ge 14$$

$$-4m - 10 + 10 \ge 14 + 10$$

$$-4m \ge 24$$

$$-\frac{1}{4}(-4m) \le -\frac{1}{4}(24)$$

$$m \le -6$$

Graphing the solution set:

49. Solving the inequality:

$$3 - 4(x-2) \le -5x + 6$$

$$3 - 4x + 8 \le -5x + 6$$

$$-4x + 11 \le -5x + 6$$

$$-4x + 5x + 11 \le -5x + 5x + 6$$

$$x + 11 \le 6$$

$$x + 11 + (-11) \le 6 + (-11)$$

$$x \le -5$$

Graphing the solution set:

51. Solving for y:

$$3x + 2y < 6$$

$$2y < -3x + 6$$

$$y < -\frac{3}{2}x + 3$$

53. Solving for y:

$$2x - 5y > 10$$

$$-5y > -2x + 10$$

$$y < \frac{2}{5}x - 2$$

55. Solving for y:
$$-3x + 7y \le 21$$
$$7y \le 3x + 21$$
$$y \le \frac{3}{7}x + 3$$

57. Solving for y:
$$2x - 4y \ge -4$$
$$-4y \ge -2x - 4$$
$$y \le \frac{1}{2}x + 1$$

59. **a.** Evaluating when $x = 0$: $-5x + 3 = -5(0) + 3 = 0 + 3 = 3$

   b. Solving the equation:
$$-5x + 3 = -7$$
$$-5x + 3 - 3 = -7 - 3$$
$$-5x = -10$$
$$-\frac{1}{5}(-5x) = -\frac{1}{5}(-10)$$
$$x = 2$$

   c. Substituting $x = 0$:
$$-5(0) + 3 < -7$$
$$3 < -7 \quad \text{(false)}$$
No, $x = 0$ is not a solution to the inequality.

   d. Solving the inequality:
$$-5x + 3 < -7$$
$$-5x + 3 - 3 < -7 - 3$$
$$-5x < -10$$
$$-\frac{1}{5}(-5x) > -\frac{1}{5}(-10)$$
$$x > 2$$

61. The inequality is $x < 3$.

63. The inequality is $x < 3$.

65. Let x and $x + 1$ represent the integers. Solving the inequality:
$$x + x + 1 \ge 583$$
$$2x + 1 \ge 583$$
$$2x \ge 582$$
$$x \ge 291$$
The two numbers are at least 291.

67. Let x represent the number. Solving the inequality:
$$2x + 6 < 10$$
$$2x < 4$$
$$x < 2$$

69. Let x represent the number. Solving the inequality:
$$4x > x - 8$$
$$3x > -8$$
$$x > -\frac{8}{3}$$

71. Let w represent the width, so $3w$ represents the length. Using the formula for perimeter:
$$2(w) + 2(3w) \ge 48$$
$$2w + 6w \ge 48$$
$$8w \ge 48$$
$$w \ge 6$$
The width is at least 6 meters.

73. Let x, $x + 2$, and $x + 4$ represent the sides of the triangle. The inequality is:
$$x + (x + 2) + (x + 4) > 24$$
$$3x + 6 > 24$$
$$3x > 18$$
$$x > 6$$
The shortest side is an even number greater than 6 inches (greater than or equal to 8 inches).

75. Solving the inequality:
$$2x - 1 \geq 3$$
$$2x \geq 4$$
$$x \geq 2$$

77. Solving the inequality:
$$-2x > -8$$
$$x < 4$$

79. Solving the inequality:
$$-3 \leq 4x + 1$$
$$4x \geq -4$$
$$x \geq -1$$

2.9 Compound Inequalities

1. Graphing the solution set:

(open circle at -1, open circle at 5, shaded outside)

3. Graphing the solution set:

(open circle at -3, closed circle at 0)

5. Graphing the solution set:

(open circle at -1, closed circle at 6)

7. Graphing the solution set:

(open circle at 2, open circle at 4)

9. Graphing the solution set:

(closed circle at -2, closed circle at 4)

11. Graphing the solution set:

(open circle at -1, open circle at 5)

13. Graphing the solution set:

(open circle at -1, open circle at 3)

15. Graphing the solution set:

(open circle at -3, closed circle at -2)

17. Solving the compound inequality:
$$3x - 1 < 5 \qquad \text{or} \qquad 5x - 5 > 10$$
$$3x < 6 \qquad\qquad\qquad 5x > 15$$
$$x < 2 \qquad\qquad\qquad x > 3$$

Graphing the solution set:

(open circle at 2, open circle at 3)

19. Solving the compound inequality:
$$x - 2 > -5 \qquad \text{and} \qquad x + 7 < 13$$
$$x > -3 \qquad\qquad\qquad x < 6$$

Graphing the solution set:

(open circle at -3, open circle at 6)

21. Solving the compound inequality:
$$11x < 22 \qquad \text{or} \qquad 12x > 36$$
$$x < 2 \qquad\qquad\qquad x > 3$$

Graphing the solution set:

(open circle at 2, open circle at 3)

23. Solving the compound inequality:
$$3x - 5 < 10 \qquad \text{and} \qquad 2x + 1 > -5$$
$$3x < 15 \qquad\qquad\qquad 2x > -6$$
$$x < 5 \qquad\qquad\qquad x > -3$$

Graphing the solution set:

(open circle at -3, open circle at 5)

25. Solving the compound inequality:
$$2x - 3 < 8 \qquad \text{and} \qquad 3x + 1 > -10$$
$$2x < 11 \qquad\qquad\qquad 3x > -11$$
$$x < \frac{11}{2} \qquad\qquad\qquad x > -\frac{11}{3}$$

Graphing the solution set:

$-11/3$ $11/2$

27. Solving the compound inequality:
$$2x - 1 < 3 \qquad \text{and} \qquad 3x - 2 > 1$$
$$2x < 4 \qquad\qquad\qquad 3x > 3$$
$$x < 2 \qquad\qquad\qquad x > 1$$

Graphing the solution set:

1 2

29. Solving the compound inequality:
$$-1 \le x - 5 \le 2$$
$$4 \le x \le 7$$

Graphing the solution set:

4 7

31. Solving the compound inequality:
$$-4 \le 2x \le 6$$
$$-2 \le x \le 3$$

Graphing the solution set:

-2 3

33. Solving the compound inequality:
$$-3 < 2x + 1 < 5$$
$$-4 < 2x < 4$$
$$-2 < x < 2$$

Graphing the solution set:

-2 2

35. Solving the compound inequality:
$$0 \le 3x + 2 \le 7$$
$$-2 \le 3x \le 5$$
$$-\frac{2}{3} \le x \le \frac{5}{3}$$

Graphing the solution set:

$-2/3$ $5/3$

37. Solving the compound inequality:
$$-7 < 2x + 3 < 11$$
$$-10 < 2x < 8$$
$$-5 < x < 4$$

Graphing the solution set:

-5 4

39. Solving the compound inequality:
$$-1 \le 4x + 5 \le 9$$
$$-6 \le 4x \le 4$$
$$-\frac{3}{2} \le x \le 1$$

Graphing the solution set:

$-3/2$ 1

41. The inequality is $-2 < x < 3$.

43. The inequality is $x \le -2$ or $x \ge 3$.

45. **a.** The three inequalities are $2x + x > 10$, $2x + 10 > x$, and $x + 10 > 2x$.

     b. For $x + 10 > 2x$ we have $x < 10$. For $3x > 10$ we have $x > \dfrac{10}{3}$.

        The compound inequality is $\dfrac{10}{3} < x < 10$.

47. Graphing the inequality:

50 266

49. Let x represent the number. Solving the inequality:
$$5 < 2x - 3 < 7$$
$$8 < 2x < 10$$
$$4 < x < 5$$
The number is between 4 and 5.

51. a.   The perimeter inequality is given by $20 < P < 50$.
   b.   Let w represent the width and $w + 4$ represent the length. Using the perimeter formula:

$$20 < 2w + 2(w + 4) < 30$$
$$20 < 2w + 2w + 8 < 30$$
$$20 < 4w + 8 < 30$$
$$12 < 4w < 22$$
$$3 < w < \frac{11}{2}$$

   c.   Since the length is $w + 4$, the length is given by:

$$3 + 4 < l < \frac{11}{2} + 4$$
$$7 < l < \frac{19}{2}$$

53. Finding the number: $0.25(32) = 8$

55. Finding the number: $0.20(120) = 24$

57. Finding the percent:
$$p \bullet 36 = 9$$
$$p = \frac{9}{36} = 0.25 = 25\%$$

59. Finding the percent:
$$p \bullet 50 = 5$$
$$p = \frac{5}{50} = 0.10 = 10\%$$

61. Finding the number:
$$0.20n = 16$$
$$n = \frac{16}{0.20} = 80$$

63. Finding the number:
$$0.02n = 8$$
$$n = \frac{8}{0.02} = 400$$

65. Simplifying the expression: $-|-5| = -(5) = -5$

67. Simplifying the expression: $-3 - 4(-2) = -3 + 8 = 5$

69. Simplifying the expression: $5|3 - 8| - 6|2 - 5| = 5|-5| - 6|-3| = 5(5) - 6(3) = 25 - 18 = 7$

71. Simplifying the expression:
$$5 - 2\big[-3(5 - 7) - 8\big] = 5 - 2\big[-3(-2) - 8\big] = 5 - 2(6 - 8) = 5 - 2(-2) = 5 + 4 = 9$$

73. The expression is: $-3 - (-9) = -3 + 9 = 6$

75. Applying the distributive property: $\frac{1}{2}(4x - 6) = \frac{1}{2} \bullet 4x - \frac{1}{2} \bullet 6 = 2x - 3$

77. The integers are: $-3, 0, 2$

Chapter 2 Test

See www.mathtv.com for video solutions to all problems in this chapter test.

Chapter 3
Linear Equations and Inequalities in Two Variables

3.1 Paired Data and Graphing Ordered Pairs

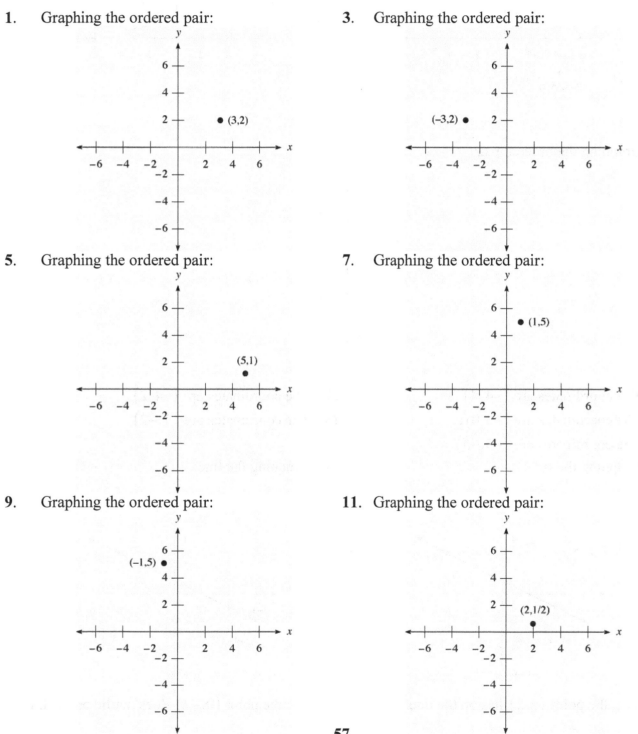

1. Graphing the ordered pair:

   • (3,2)

3. Graphing the ordered pair:

   (−3,2) •

5. Graphing the ordered pair:

   (5,1)
   •

7. Graphing the ordered pair:

   • (1,5)

9. Graphing the ordered pair:

   (−1,5) •

11. Graphing the ordered pair:

   (2,1/2)
   •

13.   Graphing the ordered pair:

15.   Graphing the ordered pair:

17.   Graphing the ordered pair:

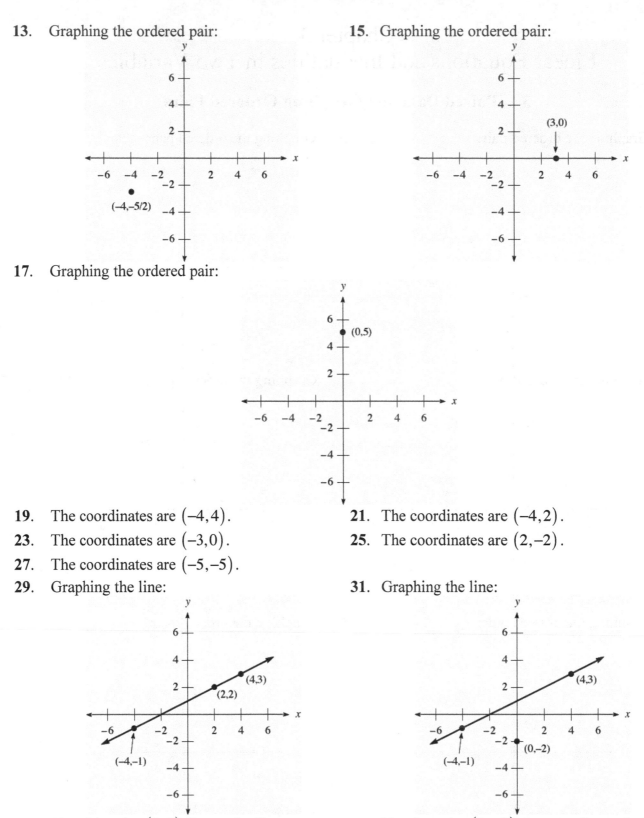

19.   The coordinates are $(-4,4)$.

21.   The coordinates are $(-4,2)$.

23.   The coordinates are $(-3,0)$.

25.   The coordinates are $(2,-2)$.

27.   The coordinates are $(-5,-5)$.

29.   Graphing the line:

31.   Graphing the line:

Yes, the point $(2,2)$ lies on the line.

No, the point $(0,-2)$ does not lie on the line.

33. Graphing the line:

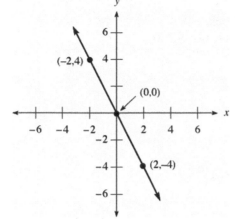

Yes, the point $(0,0)$ lies on the line.

35. Graphing the line:

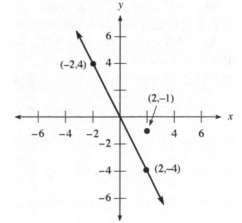

No, the point $(2,-1)$ does not lie on the line.

37. Graphing the line:

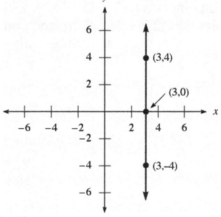

Yes, the point $(3,0)$ lies on the line.

39. No, the x-coordinate of every point on this line is 3.

41. Graphing the line:

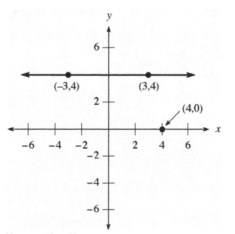

No, the point $(4,0)$ does not lie on the line.

43. No, the y-coordinate of every point on this line is 4.

45. **a.**   Three ordered pairs on the graph are (5, 40), (10, 80), and (20, 160).
      b.   She will earn $320 for working 40 hours.
      c.   If her check is $240, she worked 30 hours that week.
      d.   No. She should be paid $280 for working 35 hours, not $260.

47. Constructing a line graph:

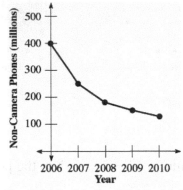

49. Five ordered pairs are (1985, 20.2), (1990, 34.4), (1995, 44.8), (2000, 65.4), and (2005, 104).
51. Point A is (6 − 5,2) = (1,2), and point B is (6,2 + 5) = (6,7).
53. Point A is (7 − 5,2) = (2,2), point B is (2,2 + 3) = (2,5), and point C is (2 + 5,5) = (7,5).
55. **a.**   Substituting $y = 4$:

$$2x + 3(4) = 6$$
$$2x + 12 = 6$$
$$2x = -6$$
$$x = -3$$

      b.   Substituting $y = -2$:

$$2x + 3(-2) = 6$$
$$2x - 6 = 6$$
$$2x = 12$$
$$x = 6$$

      c.   Substituting $x = 3$:

$$2(3) + 3y = 6$$
$$6 + 3y = 6$$
$$3y = 0$$
$$y = 0$$

      d.   Substituting $x = 9$:

$$2(9) + 3y = 6$$
$$18 + 3y = 6$$
$$3y = -12$$
$$y = -4$$

57. **a.**   Substituting $y = 7$:

$$2x - 1 = 7$$
$$2x = 8$$
$$x = 4$$

      b.   Substituting $y = 3$:

$$2x - 1 = 3$$
$$2x = 4$$
$$x = 2$$

      c.   Substituting $x = 0$: $y = 2(0) - 1 = 0 - 1 = -1$

      d.   Substituting $x = 5$: $y = 2(5) - 1 = 10 - 1 = 9$

3.2 Solutions to Linear Equations in Two Variables

1. Substituting $x = 0$, $y = 0$, and $y = -6$:

$$2(0)+y=6 \qquad 2x+0=6 \qquad 2x+(-6)=6$$
$$0+y=6 \qquad 2x=6 \qquad 2x=12$$
$$y=6 \qquad x=3 \qquad x=6$$

The ordered pairs are $(0,6)$, $(3,0)$, and $(6,-6)$.

3. Substituting $x = 0$, $y = 0$, and $x = -4$:

$$3(0)+4y=12 \qquad 3x+4(0)=12 \qquad 3(-4)+4y=12$$
$$0+4y=12 \qquad 3x+0=12 \qquad -12+4y=12$$
$$4y=12 \qquad 3x=12 \qquad 4y=24$$
$$y=3 \qquad x=4 \qquad y=6$$

The ordered pairs are $(0,3)$, $(4,0)$, and $(-4,6)$.

5. Substituting $x = 1$, $y = 0$, and $x = 5$:

$$y=4(1)-3 \qquad 0=4x-3 \qquad y=4(5)-3$$
$$y=4-3 \qquad 3=4x \qquad y=20-3$$
$$y=1 \qquad x=\tfrac{3}{4} \qquad y=17$$

The ordered pairs are $(1,1)$, $\left(\dfrac{3}{4},0\right)$, and $(5,17)$.

7. Substituting $x = 2$, $y = 6$, and $x = 0$:

$$y=7(2)-1 \qquad 6=7x-1 \qquad y=7(0)-1$$
$$y=14-1 \qquad 7=7x \qquad y=0-1$$
$$y=13 \qquad x=1 \qquad y=-1$$

The ordered pairs are $(2,13)$, $(1,6)$, and $(0,-1)$.

9. Substituting $y = 4$, $y = -3$, and $y = 0$ results (in each case) in $x = -5$. The ordered pairs are $(-5,4)$, $(-5,-3)$, and $(-5,0)$.

11. Completing the table:

x	y
1	3
-3	-9
4	12
6	18

13. Completing the table:

x	y
0	0
-1/2	-2
-3	-12
3	12

15. Completing the table:

x	y
2	3
3	2
5	0
9	-4

17. Completing the table:

x	y
2	0
3	2
1	-2
-3	-10

19. Completing the table:

x	y
0	−1
−1	−7
−3	−19
3/2	8

21. Substituting each ordered pair into the equation:

$(2,3)$: $2(2)-5(3)=4-15=-11\neq10$

$(0,-2)$: $2(0)-5(-2)=0+10=10$

$\left(\dfrac{5}{2},1\right)$: $2\left(\dfrac{5}{2}\right)-5(1)=5-5=0\neq10$

Only the ordered pair $(0,-2)$ is a solution.

23. Substituting each ordered pair into the equation:

$(1,5)$: $7(1)-2=7-2=5$

$(0,-2)$: $7(0)-2=0-2=-2$

$(-2,-16)$: $7(-2)-2=-14-2=-16$

All the ordered pairs $(1,5)$, $(0,-2)$ and $(-2,-16)$ are solutions.

25. Substituting each ordered pair into the equation:

$(1,6)$: $6(1)=6$

$(-2,12)$: $6(-2)=-12$

$(0,0)$: $6(0)=0$

The ordered pairs $(1,6)$ and $(0,0)$ are solutions.

27. Substituting each ordered pair into the equation:

$(1,1)$: $1+1=2\neq0$

$(2,-2)$: $2+(-2)=0$

$(3,3)$: $3+3=6\neq0$

Only the ordered pair $(2,-2)$ is a solution.

29. Since $x=3$, the ordered pair $(5,3)$ cannot be a solution. The ordered pairs $(3,0)$ and $(3,-3)$ are solutions.

31. Substituting $w=3$:

$2l+2(3)=30$

$2l+6=30$

$2l=24$

$l=12$

The length is 12 inches.

33. a. This is correct, since; $y=12(5)=\$60$

   b. This is not correct, since: $y=12(9)=\$108$. Her check should be for $108.

   c. This is not correct, since: $y=12(7)=\$84$. Her check should be for $84.

   d. This is correct, since: $y=12(14)=\$168$

35. **a.** Substituting $t = 5$: $V = -45,000(5) + 600,000 = -225,000 + 600,000 = \$375,000$

b. Solving when $V = 330,000$:

$$-45,000t + 600,000 = 330,000$$
$$-45,000t = -270,000$$
$$t = 6$$

The crane will be worth $330,000 at the end of 6 years.

c. Substituting $t = 9$: $V = -45,000(9) + 600,000 = -405,000 + 600,000 = \$195,000$

No, the crane will be worth $195,000 after 9 years.

d. The crane cost $600,000 (the value when $t = 0$).

37. Substituting $x = 4$:

$$3(4) + 2y = 6$$
$$12 + 2y = 6$$
$$2y = -6$$
$$y = -3$$

39. Substituting $x = 0$: $y = -\dfrac{1}{3}(0) + 2 = 0 + 2 = 2$ **41.** Substituting $x = 2$: $y = \dfrac{3}{2}(2) - 3 = 3 - 3 = 0$

43. Solving for y:

$$5x + y = 4$$
$$y = -5x + 4$$

45. Solving for y:

$$3x - 2y = 6$$
$$-2y = -3x + 6$$
$$y = \frac{3}{2}x - 3$$

3.3 Graphing Linear Equations in Two Variables

1. The ordered pairs are $(0,4)$, $(2,2)$, and $(4,0)$:

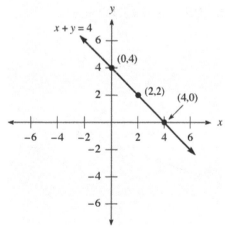

3.   The ordered pairs are $(0,3)$, $(2,1)$, and $(4,-1)$:

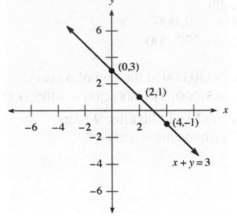

5.   The ordered pairs are $(0,0)$, $(-2,-4)$, and $(2,4)$:

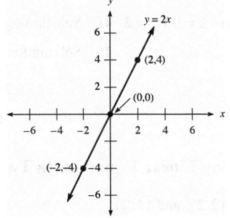

7.   The ordered pairs are $(-3,-1)$, $(0,0)$, and $(3,1)$:

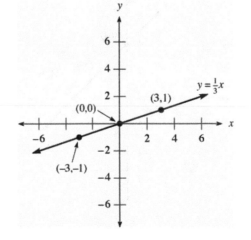

9.   The ordered pairs are $(0,1)$, $(-1,-1)$, and $(1,3)$:

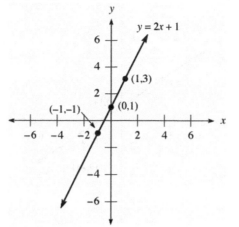

11.   The ordered pairs are $(0,4)$, $(-1,4)$, and $(2,4)$:

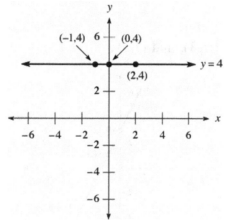

13.   The ordered pairs are $(-2,2)$, $(0,3)$, and $(2,4)$:

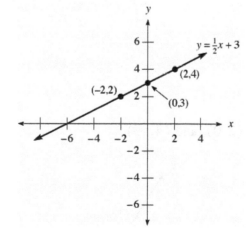

15.   The ordered pairs are $(-3,3)$, $(0,1)$, and $(3,-1)$:

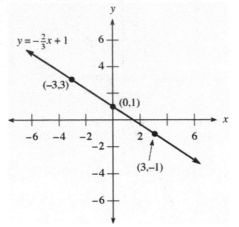

17.   Solving for y:
$$2x + y = 3$$
$$y = -2x + 3$$

The ordered pairs are $(-1,5)$, $(0,3)$, and $(1,1)$:

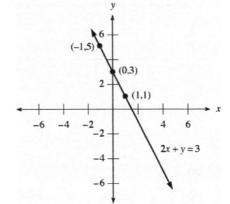

19.   Solving for y:
$$3x + 2y = 6$$
$$2y = -3x + 6$$
$$y = -\frac{3}{2}x + 3$$

The ordered pairs are $(0,3)$, $(2,0)$, and $(4,-3)$:

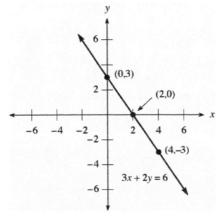

21. Solving for y:
$$-x + 2y = 6$$
$$2y = x + 6$$
$$y = \frac{1}{2}x + 3$$
The ordered pairs are $(-2,2)$, $(0,3)$, and $(2,4)$:

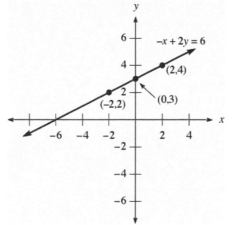

23. Three solutions are $(-4,2)$, $(0,0)$, and $(4,-2)$:

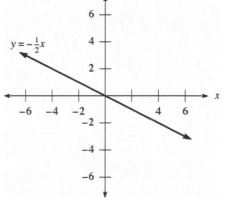

25. Three solutions are $(-1,-4)$, $(0,-1)$, and $(1,2)$:

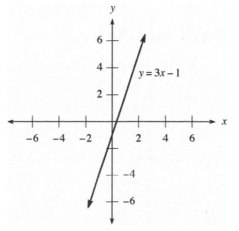

27. Solving for y:
$$-2x + y = 1$$
$$y = 2x + 1$$

Three solutions are $(-2,-3)$, $(0,1)$, and $(2,5)$:

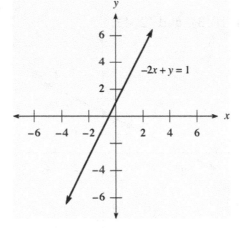

29. Solving for y:
$$3x + 4y = 8$$
$$4y = -3x + 8$$
$$y = -\frac{3}{4}x + 2$$

Three solutions are $(-4,5)$, $(0,2)$, and $(4,-1)$:

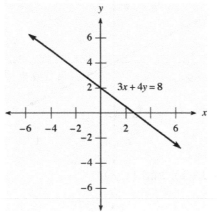

31. Three solutions are $(-2,-4)$, $(-2,0)$, and $(-2,4)$:

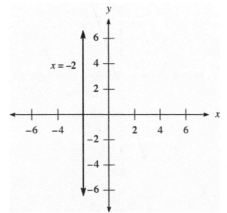

33. Three solutions are $(-4,2)$, $(0,2)$, and $(4,2)$:

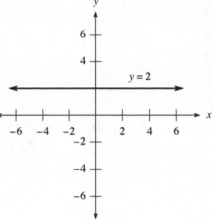

35. Graphing the equation:

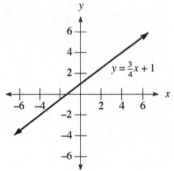

37. Graphing the equation:

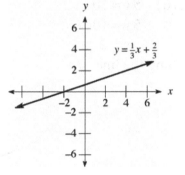

39. Graphing the equation:

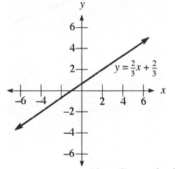

41. Completing the table:

Equation	H, V, and/or O
$x = 3$	V
$y = 3$	H
$y = 3x$	O
$y = 0$	O,H

43. Completing the table:

Equation	H, V, and/or O
$x = -\dfrac{3}{5}$	V
$y = -\dfrac{3}{5}$	H
$y = -\dfrac{3}{5}x$	O
$x = 0$	O,V

45.   Completing the table:

x	y
-4	-3
-2	-2
0	-1
2	0
6	2

47.   **a.**   Solving the equation:
$$2x + 5 = 10$$
$$2x = 5$$
$$x = \frac{5}{2}$$

b.   Substituting $y = 0$:
$$2x + 5(0) = 10$$
$$2x = 10$$
$$x = 5$$

c.   Substituting $x = 0$:
$$2(0) + 5y = 10$$
$$5y = 10$$
$$y = 2$$

d.   Graphing the line:

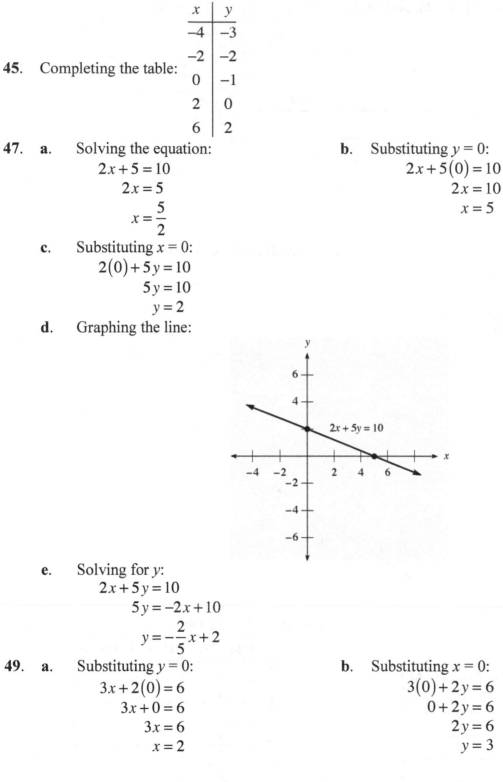

e.   Solving for y:
$$2x + 5y = 10$$
$$5y = -2x + 10$$
$$y = -\frac{2}{5}x + 2$$

49.   **a.**   Substituting $y = 0$:
$$3x + 2(0) = 6$$
$$3x + 0 = 6$$
$$3x = 6$$
$$x = 2$$

b.   Substituting $x = 0$:
$$3(0) + 2y = 6$$
$$0 + 2y = 6$$
$$2y = 6$$
$$y = 3$$

51. **a.**   Substituting $y = 0$:

$$-x + 2(0) = 4$$
$$-x + 0 = 4$$
$$-x = 4$$
$$x = -4$$

b.   Substituting $x = 0$:

$$-(0) + 2y = 4$$
$$0 + 2y = 4$$
$$2y = 4$$
$$y = 2$$

53. **a.**   Substituting $y = 0$:

$$0 = -\frac{1}{3}x + 2$$
$$-\frac{1}{3}x = -2$$
$$x = 6$$

b.   Substituting $x = 0$:

$$y = -\frac{1}{3}(0) + 2$$
$$y = 2$$

3.4 More on Graphing: Intercepts

1.   To find the x-intercept, let $y = 0$:

$$2x + 0 = 4$$
$$2x = 4$$
$$x = 2$$

To find the y-intercept, let $x = 0$:

$$2(0) + y = 4$$
$$0 + y = 4$$
$$y = 4$$

Graphing the line:

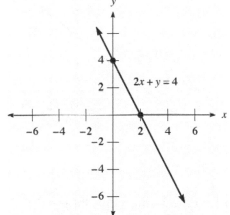

3.   To find the x-intercept, let $y = 0$:

$$-x + 0 = 3$$
$$-x = 3$$
$$x = -3$$

To find the y-intercept, let $x = 0$:

$$-0 + y = 3$$
$$y = 3$$

Graphing the line:

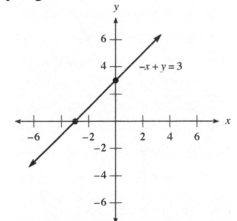

5.   To find the x-intercept, let $y = 0$:

$$-x + 2(0) = 2$$
$$-x = 2$$
$$x = -2$$

To find the y-intercept, let $x = 0$:

$$-0 + 2y = 2$$
$$2y = 2$$
$$y = 1$$

7.   To find the x-intercept, let $y = 0$:

$$5x + 2(0) = 10$$
$$5x = 10$$
$$x = 2$$

To find the y-intercept, let $x = 0$:

$$5(0) + 2y = 10$$
$$2y = 10$$
$$y = 5$$

Graphing the line: Graphing the line:

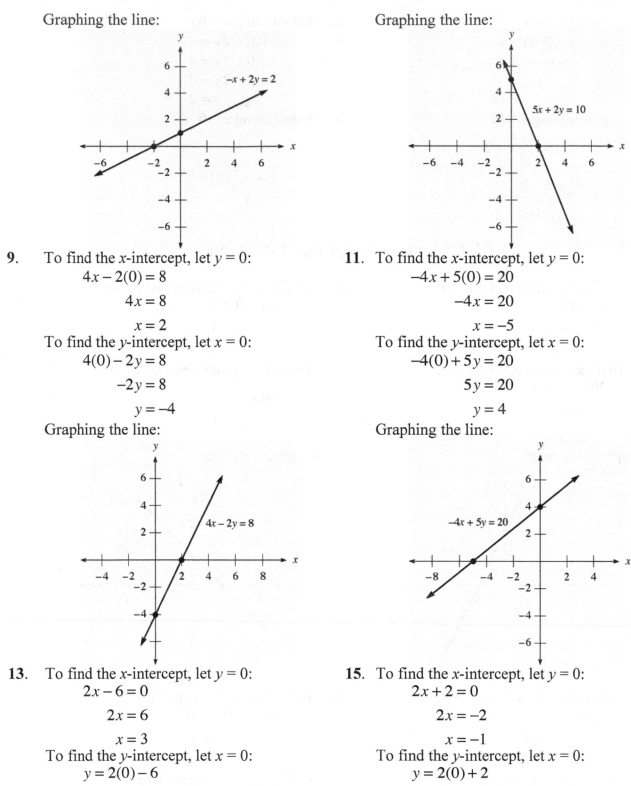

9.   To find the x-intercept, let $y = 0$:   11. To find the x-intercept, let $y = 0$:
$$4x - 2(0) = 8$$ $$-4x + 5(0) = 20$$
$$4x = 8$$ $$-4x = 20$$
$$x = 2$$ $$x = -5$$
To find the y-intercept, let $x = 0$:   To find the y-intercept, let $x = 0$:
$$4(0) - 2y = 8$$ $$-4(0) + 5y = 20$$
$$-2y = 8$$ $$5y = 20$$
$$y = -4$$ $$y = 4$$
Graphing the line: Graphing the line:

13. To find the x-intercept, let $y = 0$:   15. To find the x-intercept, let $y = 0$:
$$2x - 6 = 0$$ $$2x + 2 = 0$$
$$2x = 6$$ $$2x = -2$$
$$x = 3$$ $$x = -1$$
To find the y-intercept, let $x = 0$:   To find the y-intercept, let $x = 0$:
$$y = 2(0) - 6$$ $$y = 2(0) + 2$$
$$y = -6$$ $$y = 2$$

Graphing the line:

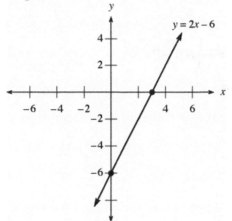

Graphing the line:

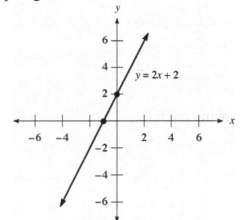

17. To find the *x*-intercept, let $y = 0$:

$$2x - 1 = 0$$
$$2x = 1$$
$$x = \frac{1}{2}$$

To find the *y*-intercept, let $x = 0$:

$$y = 2(0) - 1$$
$$y = -1$$

Graphing the line:

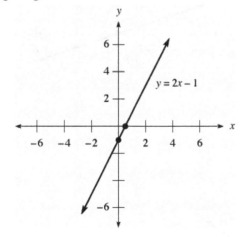

19. To find the *x*-intercept, let $y = 0$:

$$\frac{1}{2}x + 3 = 0$$
$$\frac{1}{2}x = -3$$
$$x = -6$$

To find the *y*-intercept, let $x = 0$:

$$y = \frac{1}{2}(0) + 3$$
$$y = 3$$

Graphing the line:

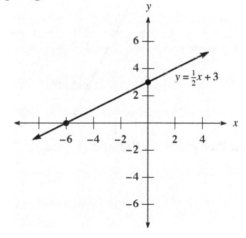

21. To find the *x*-intercept, let *y* = 0:   To find the *y*-intercept, let *x* = 0:

$$-\frac{1}{3}x - 2 = 0$$

$$-\frac{1}{3}x = 2$$

$$x = -6$$

$$y = -\frac{1}{3}(0) - 2$$

$$y = -2$$

Graphing the line:

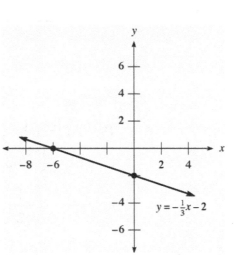

23. Another point on the line is $(2,-4)$:

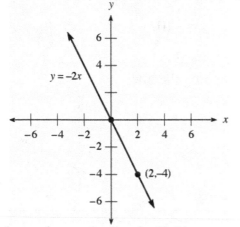

25. Another point on the line is $(3,-1)$:

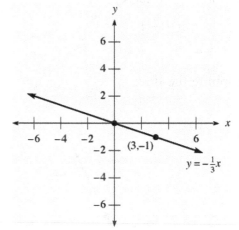

27. Another point on the line is $(3,2)$:

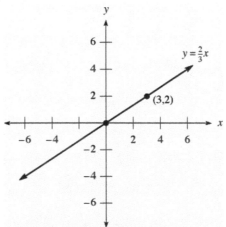

29. Completing the table:

Equation	x-intercept	y-intercept
$3x + 4y = 12$	4	3
$3x + 4y = 4$	$\frac{4}{3}$	1
$3x + 4y = 3$	1	$\frac{3}{4}$
$3x + 4y = 2$	$\frac{2}{3}$	$\frac{1}{2}$

31. Completing the table:

Equation	x-intercept	y-intercept
$x - 3y = 2$	2	$-\frac{2}{3}$
$y = \frac{1}{3}x - \frac{2}{3}$	2	$-\frac{2}{3}$
$x - 3y = 0$	0	0
$y = \frac{1}{3}x$	0	0

33. **a.** Solving the equation:

$$2x - 3 = -3$$
$$2x = 0$$
$$x = 0$$

b. Substituting $y = 0$:

$$2x - 3(0) = -3$$
$$2x - 0 = -3$$
$$2x = -3$$
$$x = -\frac{3}{2}$$

c. Substituting $x = 0$:

$$2(0) - 3y = -3$$
$$0 - 3y = -3$$
$$-3y = -3$$
$$y = 1$$

d. Graphing the equation:

e. Solving for y:

$$2x - 3y = -3$$
$$-3y = -2x - 3$$
$$y = \frac{2}{3}x + 1$$

35. The x-intercept is 3 and the y-intercept is 5.   **37.** The x-intercept is –1 and the y-intercept is –3.

39.   Graphing the line:

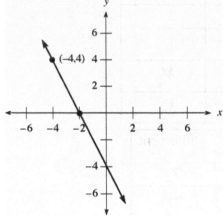

The y-intercept is -4.

41.   Graphing the line:

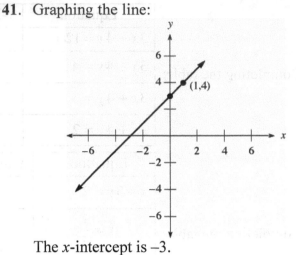

The x-intercept is -3.

43.   Graphing the line:

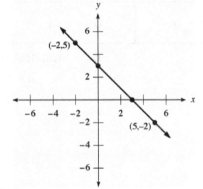

The x- and y-intercepts are both 3.

45.   Completing the table:

x	y
-2	1
0	-1
-1	0
1	-2

47.   Graphing the line:

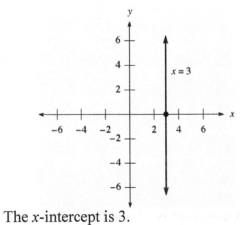

The x-intercept is 3.

49.   Graphing the line:

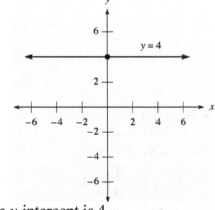

The y-intercept is 4.

51. Sketching the graph:

53. a. Evaluating: $\dfrac{5-2}{3-1} = \dfrac{3}{2}$

b. Evaluating: $\dfrac{2-5}{1-3} = \dfrac{-3}{-2} = \dfrac{3}{2}$

55. a. Evaluating when $x = 3$ and $y = 5$: $\dfrac{y-2}{x-1} = \dfrac{5-2}{3-1} = \dfrac{3}{2}$

b. Evaluating when $x = 3$ and $y = 5$: $\dfrac{2-y}{1-x} = \dfrac{2-5}{1-3} = \dfrac{-3}{-2} = \dfrac{3}{2}$

3.5 The Slope of a Line

1. The slope is given by: $m = \dfrac{4-1}{4-2} = \dfrac{3}{2}$

3. The slope is given by: $m = \dfrac{2-4}{5-1} = \dfrac{-2}{4} = -\dfrac{1}{2}$

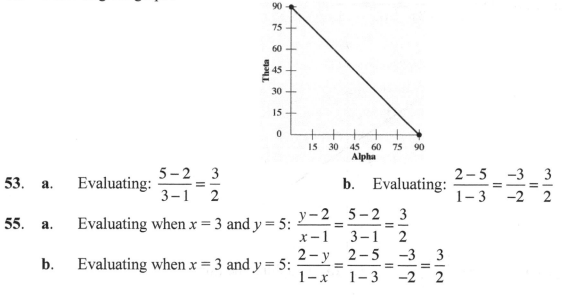

5. The slope is given by: $m = \dfrac{2-(-3)}{4-1} = \dfrac{5}{3}$

7. The slope is given by: $m = \dfrac{3-(-2)}{1-(-3)} = \dfrac{5}{4}$

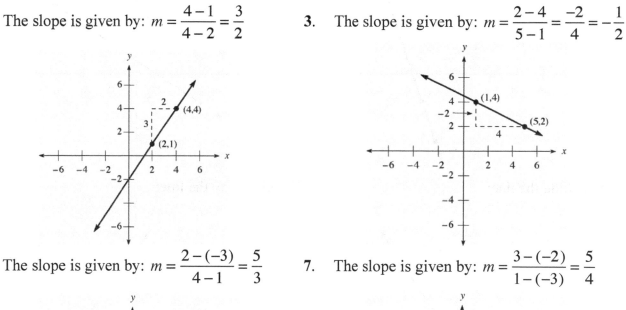

9.   The slope is given by: $m = \dfrac{-2-2}{3-(-3)} = \dfrac{-4}{6} = -\dfrac{2}{3}$

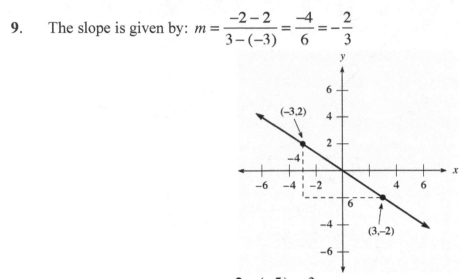

11.   The slope is given by: $m = \dfrac{-2-(-5)}{3-2} = \dfrac{3}{1} = 3$ 13. Graphing the line:

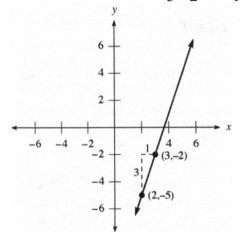

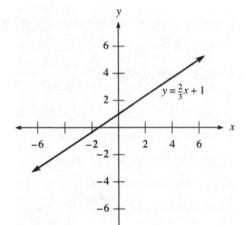

15.   Graphing the line:

17. Graphing the line:

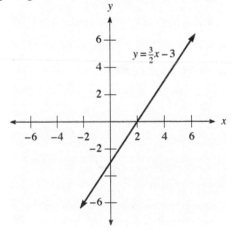

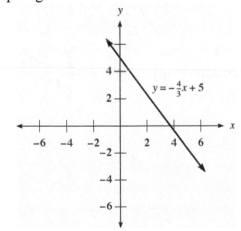

19. Graphing the line:

21. Graphing the line:

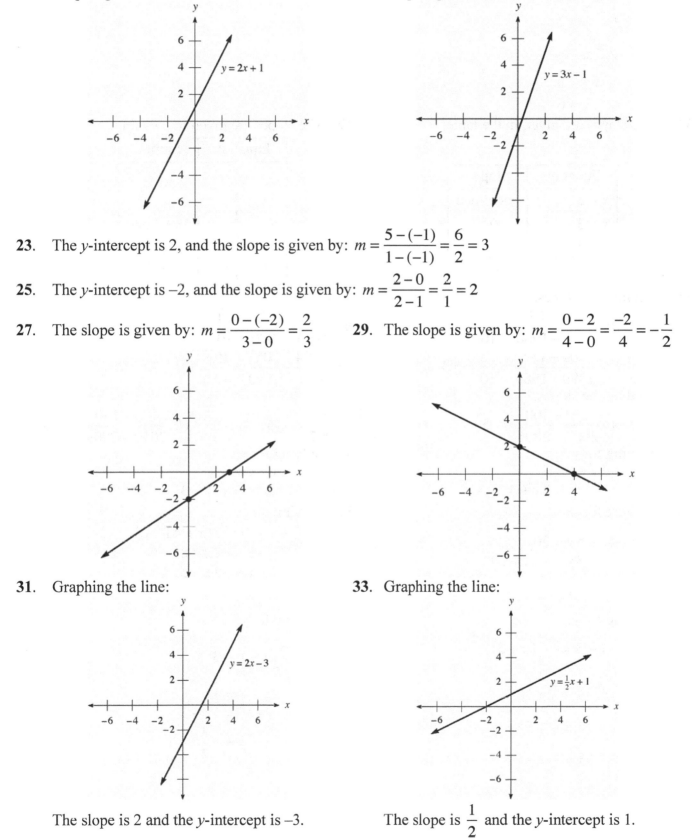

23. The y-intercept is 2, and the slope is given by: $m = \dfrac{5-(-1)}{1-(-1)} = \dfrac{6}{2} = 3$

25. The y-intercept is -2, and the slope is given by: $m = \dfrac{2-0}{2-1} = \dfrac{2}{1} = 2$

27. The slope is given by: $m = \dfrac{0-(-2)}{3-0} = \dfrac{2}{3}$

29. The slope is given by: $m = \dfrac{0-2}{4-0} = \dfrac{-2}{4} = -\dfrac{1}{2}$

31. Graphing the line:

The slope is 2 and the y-intercept is -3.

33. Graphing the line:

The slope is $\dfrac{1}{2}$ and the y-intercept is 1.

35.   Using the slope formula:

$$\frac{y-2}{6-4}=2$$

$$\frac{y-2}{2}=2$$

$$y-2=4$$

$$y=6$$

37.   The slopes are given in the table:

Equation	Slope
$x=3$	undefined
$y=3$	0
$y=3x$	3

39.   The slopes are given in the table:

Equation	Slope
$y=-\dfrac{2}{3}$	0
$x=-\dfrac{2}{3}$	undefined
$y=-\dfrac{2}{3}x$	$-\dfrac{2}{3}$

41.   Finding the slopes:

A: $\dfrac{121-88}{1970-1960}=\dfrac{33}{10}=3.3$

B: $\dfrac{152-121}{1980-1970}=\dfrac{31}{10}=3.1$

C: $\dfrac{205-152}{1990-1980}=\dfrac{53}{10}=5.3$

D: $\dfrac{224-205}{2000-1990}=\dfrac{19}{10}=1.9$

43.   Finding the slopes:

A: $\dfrac{250-300}{2007-2006}=-50$ B: $\dfrac{175-250}{2008-2007}=-75$ C: $\dfrac{125-150}{2010-2009}=-25$

45.   Solving for y:

$$-2x+y=4$$

$$y=2x+4$$

47.   Solving for y:

$$2x+y=3$$

$$y=-2x+3$$

49.   Solving for y:

$$4x-5y=20$$

$$-5y=-4x+20$$

$$y=\frac{4}{5}x-4$$

51.   Solving for y:

$$-y-3=-2(x+4)$$

$$-y-3=-2x-8$$

$$-y=-2x-5$$

$$y=2x+5$$

53.   Solving for y:

$$-y-3=-\frac{2}{3}(x+3)$$

$$-y-3=-\frac{2}{3}x-2$$

$$-y=-\frac{2}{3}x+1$$

$$y=\frac{2}{3}x-1$$

55.   Solving for y:

$$-\frac{y-1}{x}=\frac{3}{2}$$

$$-2y+2=3x$$

$$-2y=3x-2$$

$$y=-\frac{3}{2}x+1$$

3.6 Finding the Equation of a Line

1. The slope-intercept form is $y = \dfrac{2}{3}x + 1$.

3. The slope-intercept form is $y = \dfrac{3}{2}x - 1$.

5. The slope-intercept form is $y = -\dfrac{2}{5}x + 3$.

7. The slope-intercept form is $y = 2x - 4$.

9. Solving for y:

$$-2x + y = 4$$
$$y = 2x + 4$$

The slope is 2 and the y-intercept is 4:

11. Solving for y:

$$3x + y = 3$$
$$y = -3x + 3$$

The slope is -3 and the y-intercept is 3:

13. Solving for y:

$$3x + 2y = 6$$
$$2y = -3x + 6$$
$$y = -\dfrac{3}{2}x + 3$$

The slope is $-\dfrac{3}{2}$ and the y-intercept is 3:

15. Solving for y:

$$4x - 5y = 20$$
$$-5y = -4x + 20$$
$$y = \dfrac{4}{5}x - 4$$

The slope is $\dfrac{4}{5}$ and the y-intercept is -4:

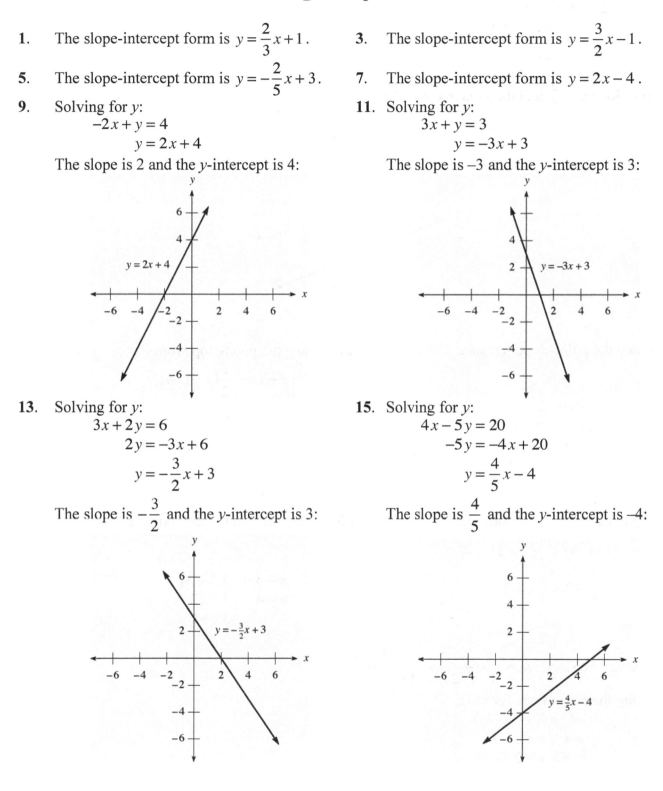

17.   Solving for y:
$$-2x - 5y = 10$$
$$-5y = 2x + 10$$
$$y = -\frac{2}{5}x - 2$$

The slope is $-\frac{2}{5}$ and the y-intercept is -2:

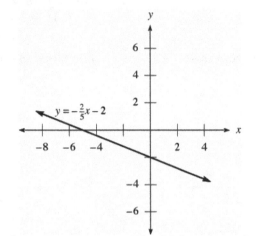

19.   Using the point-slope formula:

$$y - (-5) = 2(x - (-2))$$
$$y + 5 = 2(x + 2)$$
$$y + 5 = 2x + 4$$
$$y = 2x - 1$$

21.   Using the point-slope formula:
$$y - 1 = -\frac{1}{2}(x - (-4))$$
$$y - 1 = -\frac{1}{2}(x + 4)$$
$$y - 1 = -\frac{1}{2}x - 2$$
$$y = -\frac{1}{2}x - 1$$

23.   Using the point-slope formula:
$$y - (-3) = \frac{3}{2}(x - 2)$$
$$y + 3 = \frac{3}{2}x - 3$$
$$y = \frac{3}{2}x - 6$$

25.   Using the point-slope formula:
$$y - 4 = -3(x - (-1))$$
$$y - 4 = -3(x + 1)$$
$$y - 4 = -3x - 3$$
$$y = -3x + 1$$

27.   Finding the slope: $m = \dfrac{-1 - (-4)}{1 - (-2)} = \dfrac{-1 + 4}{1 + 2} = \dfrac{3}{3} = 1$

Using the point-slope formula:
$$y - (-4) = 1(x - (-2))$$
$$y + 4 = x + 2$$
$$y = x - 2$$

29. Finding the slope: $m = \dfrac{1-(-5)}{2-(-1)} = \dfrac{1+5}{2+1} = \dfrac{6}{3} = 2$

    Using the point-slope formula:
    $$y - 1 = 2(x - 2)$$
    $$y - 1 = 2x - 4$$
    $$y = 2x - 3$$

31. Finding the slope: $m = \dfrac{6-(-2)}{3-(-3)} = \dfrac{6+2}{3+3} = \dfrac{8}{6} = \dfrac{4}{3}$

    Using the point-slope formula:
    $$y - 6 = \dfrac{4}{3}(x - 3)$$
    $$y - 6 = \dfrac{4}{3}x - 4$$
    $$y = \dfrac{4}{3}x + 2$$

33. Finding the slope: $m = \dfrac{-5-(-1)}{3-(-3)} = \dfrac{-5+1}{3+3} = \dfrac{-4}{6} = -\dfrac{2}{3}$

    Using the point-slope formula:
    $$y - (-5) = -\dfrac{2}{3}(x - 3)$$
    $$y + 5 = -\dfrac{2}{3}x + 2$$
    $$y = -\dfrac{2}{3}x - 3$$

35. The y-intercept is 3, and the slope is: $m = \dfrac{3-0}{0-(-1)} = \dfrac{3}{1} = 3$.

    The slope-intercept form is $y = 3x + 3$.

37. The y-intercept is -1, and the slope is: $m = \dfrac{0-(-1)}{4-0} = \dfrac{0+1}{4} = \dfrac{1}{4}$.

    The slope-intercept form is $y = \dfrac{1}{4}x - 1$.

39. **a.** Solving the equation:
    $$-2x + 1 = 6$$
    $$-2x = 5$$
    $$x = -\dfrac{5}{2}$$

    b. Writing in slope-intercept form:
    $$-2x + y = 6$$
    $$y = 2x + 6$$

    c. Substituting $x = 0$:
    $$-2(0) + y = 6$$
    $$0 + y = 6$$
    $$y = 6$$

    d. Finding the slope:
    $$-2x + y = 6$$
    $$y = 2x + 6$$

    The slope is 2.

e.   Graphing the line:

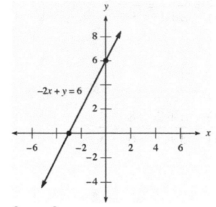

41.   The slope is given by: $m = \dfrac{0-2}{3-0} = -\dfrac{2}{3}$. Since $b = 2$, the equation is $y = -\dfrac{2}{3}x + 2$.

43.   The slope is given by: $m = \dfrac{0-(-5)}{-2-0} = -\dfrac{5}{2}$. Since $b = -5$, the equation is $y = -\dfrac{5}{2}x - 5$.

45.   Its equation is $x = 3$.

47.   **a.**   After 5 years, the copier is worth $6,000.
       b.   The copier is worth $12,000 after 3 years.
       c.   The slope of the line is −3000.
       d.   The copier is decreasing in value by $3000 per year.
       e.   The equation is $V = -3000t + 21,000$.

49.   Graphing the line: 51. Graphing the line:

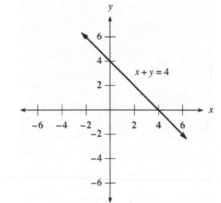
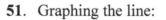
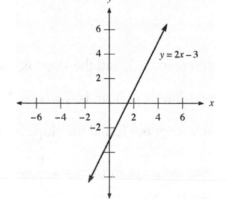

53.   Graphing the line:

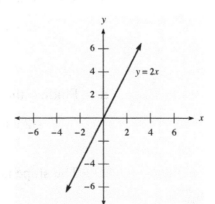

3.7 Linear Inequalities in Two Variables

1.   Checking the point $(0,0)$:

$2(0) - 3(0) = 0 - 0 < 6$ (true)

Graphing the linear inequality:

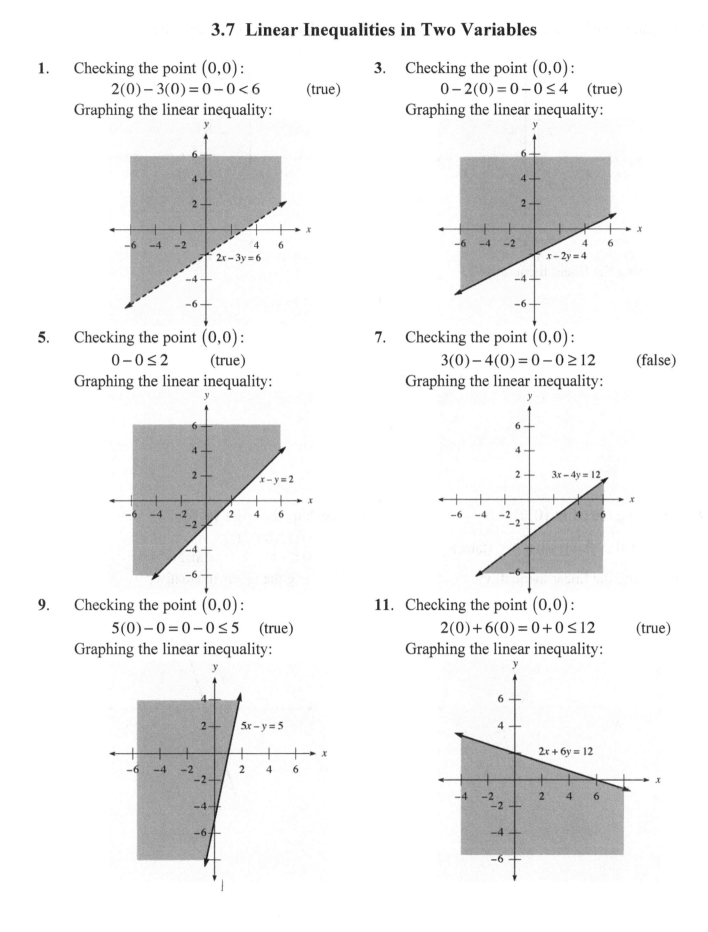

3.   Checking the point $(0,0)$:

$0 - 2(0) = 0 - 0 \le 4$ (true)

Graphing the linear inequality:

5.   Checking the point $(0,0)$:

$0 - 0 \le 2$ (true)

Graphing the linear inequality:

7.   Checking the point $(0,0)$:

$3(0) - 4(0) = 0 - 0 \ge 12$ (false)

Graphing the linear inequality:

9.   Checking the point $(0,0)$:

$5(0) - 0 = 0 - 0 \le 5$ (true)

Graphing the linear inequality:

11.   Checking the point $(0,0)$:

$2(0) + 6(0) = 0 + 0 \le 12$ (true)

Graphing the linear inequality:

13. Graphing the linear inequality:

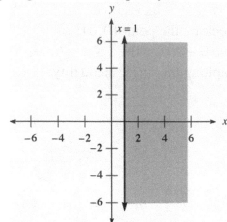

15. Graphing the linear inequality:

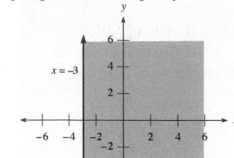

17. Graphing the linear inequality:

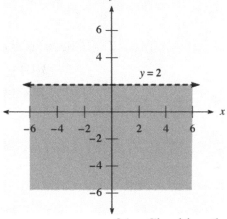

19. Checking the point $(0,0)$:

$$2(0)+0=0+0>3 \quad \text{(false)}$$

Graphing the linear inequality:

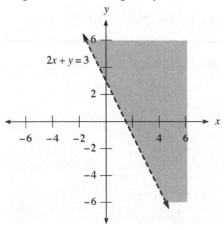

21. Checking the point $(0,0)$:

$$0 \le 3(0)-1$$
$$0 \le -1 \qquad \text{(false)}$$

Graphing the linear inequality:

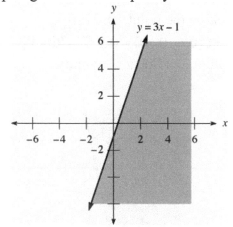

23. Checking the point $(0,0)$:

$$0 \le -\frac{1}{2}(0)+2$$
$$0 \le 2 \qquad \text{(true)}$$

Graphing the linear inequality:

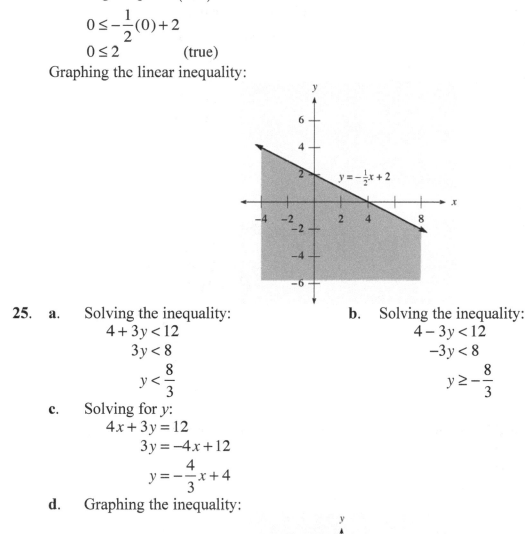

25. **a.** Solving the inequality:
$$4+3y<12$$
$$3y<8$$
$$y<\frac{8}{3}$$

b. Solving the inequality:
$$4-3y<12$$
$$-3y<8$$
$$y\ge-\frac{8}{3}$$

c. Solving for y:
$$4x+3y=12$$
$$3y=-4x+12$$
$$y=-\frac{4}{3}x+4$$

d. Graphing the inequality:

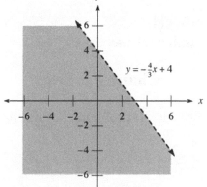

27. **a.** The slope is $\frac{2}{5}$ and the y-intercept is 2, so the equation is $y=\frac{2}{5}x+2$.

b. Since the shading is below the line, the inequality is $y<\frac{2}{5}x+2$.

c. Since the shading is above the line, the inequality is $y>\frac{2}{5}x+2$.

29.   Simplifying the expression: $7 - 3(2x - 4) - 8 = 7 - 6x + 12 - 8 = -6x + 11$

31.   Solving the equation:

$$-\frac{3}{2}x = 12$$

$$-\frac{2}{3}\left(-\frac{3}{2}x\right) = -\frac{2}{3}(12)$$

$$x = -8$$

33.   Solving the equation:

$$8 - 2(x + 7) = 2$$
$$8 - 2x - 14 = 2$$
$$-2x - 6 = 2$$
$$-2x = 8$$
$$x = -4$$

35.   Solving for w:

$$2l + 2w = P$$
$$2w = P - 2l$$
$$w = \frac{P - 2l}{2}$$

37.   Solving the inequality:

$$3 - 2x > 5$$
$$3 - 3 - 2x > 5 - 3$$
$$-2x > 2$$
$$-\frac{1}{2}(-2x) < -\frac{1}{2}(2)$$
$$x < -1$$

Graphing the solution set:

39.   Solving for y:

$$3x - 2y \le 12$$
$$3x - 3x - 2y \le -3x + 12$$
$$-2y \le -3x + 12$$
$$-\frac{1}{2}(-2y) \ge -\frac{1}{2}(-3x + 12)$$
$$y \ge \frac{3}{2}x - 6$$

41.   Let w represent the width and $3w + 5$ represent the length. Using the perimeter formula:

$$2(w) + 2(3w + 5) = 26$$
$$2w + 6w + 10 = 26$$
$$8w + 10 = 26$$
$$8w = 16$$
$$w = 2$$
$$3w + 5 = 3(2) + 5 = 11$$

The width is 2 inches and the length is 11 inches.

Chapter 3 Test

See www.mathtv.com for video solutions to all problems in this chapter test.

Chapter 4
Exponents and Polynomials

4.1 Multiplication with Exponents

1. The base is 4 and the exponent is 2. Evaluating the expression: $4^2 = 4 \cdot 4 = 16$

3. The base is 0.3 and the exponent is 2. Evaluating the expression: $0.3^2 = 0.3 \cdot 0.3 = 0.09$

5. The base is 4 and the exponent is 3. Evaluating the expression: $4^3 = 4 \cdot 4 \cdot 4 = 64$

7. The base is -5 and the exponent is 2. Evaluating the expression: $(-5)^2 = (-5) \cdot (-5) = 25$

9. The base is 2 and the exponent is 3. Evaluating the expression: $-2^3 = -2 \cdot 2 \cdot 2 = -8$

11. The base is 3 and the exponent is 4. Evaluating the expression: $3^4 = 3 \cdot 3 \cdot 3 \cdot 3 = 81$

13. The base is $\dfrac{2}{3}$ and the exponent is 2. Evaluating the expression: $\left(\dfrac{2}{3}\right)^2 = \left(\dfrac{2}{3}\right) \cdot \left(\dfrac{2}{3}\right) = \dfrac{4}{9}$

15. The base is $\dfrac{1}{2}$ and the exponent is 4. Evaluating the expression:

$$\left(\frac{1}{2}\right)^4 = \left(\frac{1}{2}\right) \cdot \left(\frac{1}{2}\right) \cdot \left(\frac{1}{2}\right) \cdot \left(\frac{1}{2}\right) = \frac{1}{16}$$

17. **a.** Completing the table:

Number (x)	1	2	3	4	5	6	7
Square (x^2)	1	4	9	16	25	36	49

   b. For numbers larger than 1, the square of the number is larger than the number.

19. Simplifying the expression: $x^4 \cdot x^5 = x^{4+5} = x^9$

21. Simplifying the expression: $y^{10} \cdot y^{20} = y^{10+20} = y^{30}$

23. Simplifying the expression: $2^5 \cdot 2^4 \cdot 2^3 = 2^{5+4+3} = 2^{12}$

25. Simplifying the expression: $x^4 \cdot x^6 \cdot x^8 \cdot x^{10} = x^{4+6+8+10} = x^{28}$

27. Simplifying the expression: $\left(x^2\right)^5 = x^{2 \cdot 5} = x^{10}$

29. Simplifying the expression: $\left(5^4\right)^3 = 5^{4 \cdot 3} = 5^{12}$

31. Simplifying the expression: $\left(y^3\right)^3 = y^{3 \cdot 3} = y^9$

33. Simplifying the expression: $\left(2^5\right)^{10} = 2^{5 \cdot 10} = 2^{50}$

35. Simplifying the expression: $\left(a^3\right)^x = a^{3x}$

37. Simplifying the expression: $\left(b^x\right)^y = b^{xy}$

39. Simplifying the expression: $(4x)^2 = 4^2 \cdot x^2 = 16x^2$

41. Simplifying the expression: $(2y)^5 = 2^5 \cdot y^5 = 32y^5$

43.   Simplifying the expression: $(-3x)^4 = (-3)^4 \cdot x^4 = 81x^4$

45.   Simplifying the expression: $(0.5ab)^2 = (0.5)^2 \cdot a^2 b^2 = 0.25 a^2 b^2$

47.   Simplifying the expression: $(4xyz)^3 = 4^3 \cdot x^3 y^3 z^3 = 64 x^3 y^3 z^3$

49.   Simplifying using properties of exponents: $(2x^4)^3 = 2^3 (x^4)^3 = 8x^{12}$

51.   Simplifying using properties of exponents: $(4a^3)^2 = 4^2 (a^3)^2 = 16a^6$

53.   Simplifying using properties of exponents: $(x^2)^3 (x^4)^2 = x^6 \cdot x^8 = x^{14}$

55.   Simplifying using properties of exponents: $(a^3)^1 (a^2)^4 = a^3 \cdot a^8 = a^{11}$

57.   Simplifying using properties of exponents: $(2x)^3 (2x)^4 = (2x)^7 = 2^7 x^7 = 128 x^7$

59.   Simplifying using properties of exponents: $(3x^2)^3 (2x)^4 = 3^3 x^6 \cdot 2^4 x^4 = 27 x^6 \cdot 16 x^4 = 432 x^{10}$

61.   Simplifying using properties of exponents: $(4x^2 y^3)^2 = 4^2 x^4 y^6 = 16 x^4 y^6$

63.   Simplifying using properties of exponents: $\left(\dfrac{2}{3} a^4 b^5\right)^3 = \left(\dfrac{2}{3}\right)^3 a^{12} b^{15} = \dfrac{8}{27} a^{12} b^{15}$

65.   **a.**   Completing the table:

Number (x)	Square (x^2)
−3	9
−2	4
−1	1
0	0
1	1
2	4
3	9

67.   Completing the table:

Number (x)	Square (x^2)
−2.5	6.25
−1.5	2.25
−0.5	0.25
0	0
0.5	0.25
1.5	2.25
2.5	6.25

b.   Constructing a line graph:

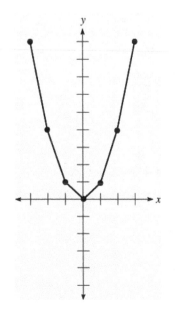

69. Writing in scientific notation: $43,200 = 4.32 \times 10^4$

71. Writing in scientific notation: $570 = 5.7 \times 10^2$

73. Writing in scientific notation: $238,000 = 2.38 \times 10^5$

75. Writing in expanded form: $2.49 \times 10^3 = 2,490$

77. Writing in expanded form: $3.52 \times 10^2 = 352$

79. Writing in expanded form: $2.8 \times 10^4 = 28,000$

81. The volume is given by: $V = (3 \text{ in.})^3 = 27 \text{ inches}^3$

83. The volume is given by: $V = (2.5 \text{ in.})^3 \approx 15.6 \text{ inches}^3$

85. The volume is given by: $V = (8 \text{ in.})(4.5 \text{ in.})(1 \text{ in.}) = 36 \text{ inches}^3$

87. Yes. If the box was 7 ft x 3 ft x 2 ft, you could fit inside of it.

89. Writing in scientific notation: $650,000,000 \text{ seconds} = 6.5 \times 10^8 \text{ seconds}$

91. Writing in expanded form: $7.4 \times 10^5 \text{ dollars} = \$740,000$

93. Writing in expanded form: $1.8 \times 10^5 \text{ dollars} = \$180,000$

95. Substitute $c = 8$, $b = 3.35$, and $s = 3.11$: $d = \pi \bullet 3.11 \bullet 8 \bullet \left(\frac{1}{2} \bullet 3.35 \right)^2 \approx 219 \text{ inches}^3$

97. Substitute $c = 6$, $b = 3.59$, and $s = 2.99$: $d = \pi \bullet 2.99 \bullet 6 \bullet \left(\frac{1}{2} \bullet 3.59 \right)^2 \approx 182 \text{ inches}^3$

99. Subtracting: $4 - 7 = 4 + (-7) = -3$

101. Subtracting: $4 - (-7) = 4 + 7 = 11$

103. Subtracting: $15 - 20 = 15 + (-20) = -5$

105. Subtracting: $-15 - (-20) = -15 + 20 = 5$

107. Simplifying: $2(3) - 4 = 6 - 4 = 2$

109. Simplifying: $4(3) - 3(2) = 12 - 6 = 6$

111. Simplifying: $2(5 - 3) = 2(2) = 4$

113. Simplifying: $5 + 4(-2) - 2(-3) = 5 - 8 + 6 = 3$

4.2 Division with Exponents

1. Writing with positive exponents: $3^{-2} = \dfrac{1}{3^2} = \dfrac{1}{9}$

3. Writing with positive exponents: $6^{-2} = \dfrac{1}{6^2} = \dfrac{1}{36}$

5. Writing with positive exponents: $8^{-2} = \dfrac{1}{8^2} = \dfrac{1}{64}$

7. Writing with positive exponents: $5^{-3} = \dfrac{1}{5^3} = \dfrac{1}{125}$

9. Writing with positive exponents: $2x^{-3} = 2 \bullet \dfrac{1}{x^3} = \dfrac{2}{x^3}$

11. Writing with positive exponents: $(2x)^{-3} = \dfrac{1}{(2x)^3} = \dfrac{1}{8x^3}$

13. Writing with positive exponents: $(5y)^{-2} = \dfrac{1}{(5y)^2} = \dfrac{1}{25y^2}$

15. Writing with positive exponents: $10^{-2} = \dfrac{1}{10^2} = \dfrac{1}{100}$

17. Completing the table:

Number (x)	Square (x^2)	Power of 2 (2^x)
-3	9	$\frac{1}{8}$
-2	4	$\frac{1}{4}$
-1	1	$\frac{1}{2}$
0	0	1
1	1	2
2	4	4
3	9	8

19. Simplifying: $\dfrac{5^1}{5^3} = 5^{1-3} = 5^{-2} = \dfrac{1}{5^2} = \dfrac{1}{25}$ **21.** Simplifying: $\dfrac{x^{10}}{x^4} = x^{10-4} = x^6$

23. Simplifying: $\dfrac{4^3}{4^0} = 4^{3-0} = 4^3 = 64$

25. Simplifying: $\dfrac{(2x)^7}{(2x)^4} = (2x)^{7-4} = (2x)^3 = 2^3 x^3 = 8x^3$

27. Simplifying: $\dfrac{6^{11}}{6} = \dfrac{6^{11}}{6^1} = 6^{11-1} = 6^{10} \quad (= 60,466,176)$

29. Simplifying: $\dfrac{6}{6^{11}} = \dfrac{6^1}{6^{11}} = 6^{1-11} = 6^{-10} = \dfrac{1}{6^{10}} \quad \left(= \dfrac{1}{60,466,176}\right)$

31. Simplifying: $\dfrac{2^{-5}}{2^3} = 2^{-5-3} = 2^{-8} = \dfrac{1}{2^8} = \dfrac{1}{256}$ **33.** Simplifying: $\dfrac{2^5}{2^{-3}} = 2^{5-(-3)} = 2^{5+3} = 2^8 = 256$

35. Simplifying: $\dfrac{(3x)^{-5}}{(3x)^{-8}} = (3x)^{-5-(-8)} = (3x)^{-5+8} = (3x)^3 = 3^3 x^3 = 27x^3$

37. Simplifying: $(3xy)^4 = 3^4 x^4 y^4 = 81x^4 y^4$ **39.** Simplifying: $10^0 = 1$

41. Simplifying: $\left(2a^2 b\right)^1 = 2a^2 b$ **43.** Simplifying: $\left(7y^3\right)^{-2} = \dfrac{1}{\left(7y^3\right)^2} = \dfrac{1}{49y^6}$

45. Simplifying: $x^{-3} \bullet x^{-5} = x^{-3-5} = x^{-8} = \dfrac{1}{x^8}$ **47.** Simplifying: $y^7 \bullet y^{-10} = y^{7-10} = y^{-3} = \dfrac{1}{y^3}$

49. Simplifying: $\dfrac{\left(x^2\right)^3}{x^4} = \dfrac{x^6}{x^4} = x^{6-4} = x^2$ **51.** Simplifying: $\dfrac{\left(a^4\right)^3}{\left(a^3\right)^2} = \dfrac{a^{12}}{a^6} = a^{12-6} = a^6$

53. Simplifying: $\dfrac{y^7}{\left(y^2\right)^8} = \dfrac{y^7}{y^{16}} = y^{7-16} = y^{-9} = \dfrac{1}{y^9}$ **55.** Simplifying: $\left(\dfrac{y^7}{y^2}\right)^8 = \left(y^{7-2}\right)^8 = \left(y^5\right)^8 = y^{40}$

57. Simplifying: $\dfrac{\left(x^{-2}\right)^3}{x^{-5}} = \dfrac{x^{-6}}{x^{-5}} = x^{-6-(-5)} = x^{-6+5} = x^{-1} = \dfrac{1}{x}$

59. Simplifying: $\left(\dfrac{x^{-2}}{x^{-5}}\right)^3 = \left(x^{-2+5}\right)^3 = \left(x^3\right)^3 = x^9$

61. Simplifying: $\dfrac{\left(a^3\right)^2\left(a^4\right)^5}{\left(a^5\right)^2} = \dfrac{a^6 \bullet a^{20}}{a^{10}} = \dfrac{a^{26}}{a^{10}} = a^{26-10} = a^{16}$

63. Simplifying: $\dfrac{\left(a^{-2}\right)^3\left(a^4\right)^2}{\left(a^{-3}\right)^{-2}} = \dfrac{a^{-6} \bullet a^8}{a^6} = \dfrac{a^2}{a^6} = a^{2-6} = a^{-4} = \dfrac{1}{a^4}$

65. Completing the table:

Number (x)	Power of 2 $\left(2^x\right)$
-3	$\frac{1}{8}$
-2	$\frac{1}{4}$
-1	$\frac{1}{2}$
0	1
1	2
2	4
3	8

Constructing the line graph:

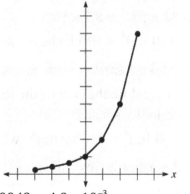

67. Writing in scientific notation: $0.0048 = 4.8 \times 10^{-3}$

69. Writing in scientific notation: $25 = 2.5 \times 10^1$

71. Writing in scientific notation: $0.000009 = 9 \times 10^{-6}$

Expanded Form	Scientific Notation $(n \times 10^r)$
0.000357	3.57×10^{-4}
0.00357	3.57×10^{-3}
0.0357	3.57×10^{-2}
0.357	3.57×10^{-1}
3.57	3.57×10^{0}
35.7	3.57×10^{1}
357	3.57×10^{2}
3,570	3.57×10^{3}
35,700	3.57×10^{4}

73. Completing the table:

75. Writing in expanded form: $4.23 \times 10^{-3} = 0.00423$

77. Writing in expanded form: $8 \times 10^{-5} = 0.00008$

79. Writing in expanded form: $4.2 \times 10^{0} = 4.2$

81. Writing in expanded form: 2×10^{-3} seconds $= 0.002$ seconds

83. Writing each number in scientific notation:
$0.002 = 2 \times 10^{-3}$, $0.005 = 5 \times 10^{-3}$, $0.006 = 6 \times 10^{-3}$,
$0.01 = 1 \times 10^{-2}$, $0.01 = 1 \times 10^{-2}$

85. Writing in scientific notation: $25 \times 10^{3} = 2.5 \times 10^{4}$

87. Writing in scientific notation: $23.5 \times 10^{4} = 2.35 \times 10^{5}$

89. Writing in scientific notation: $0.82 \times 10^{-3} = 8.2 \times 10^{-4}$

91. The area of the smaller square is $(10 \text{ in.})^2 = 100$ inches^2, while the area of the larger square is $(20 \text{ in.})^2 = 400$ inches^2. It would take 4 smaller squares to cover the larger square.

93. The area of the smaller square is x^2, while the area of the larger square is $(2x)^2 = 4x^2$. It would take 4 smaller squares to cover the larger square.

95. The volume of the smaller box is $(6 \text{ in.})^3 = 216$ inches^3, while the volume of the larger box is $(12 \text{ in.})^3 = 1,728$ inches^3. Thus 8 smaller boxes will fit inside the larger box ($8 \cdot 216 = 1,728$).

97. The volume of the smaller box is x^3, while the volume of the larger box is $(2x)^3 = 8x^3$. Thus 8 smaller boxes will fit inside the larger box.

99. Simplifying: $3(4.5) = 13.5$

101. Simplifying: $\frac{4}{5}(10) = \frac{40}{5} = 8$

103. Simplifying: $6.8(3.9) = 26.52$

105. Simplifying: $-3 + 15 = 12$

107. Simplifying: $x^5 \cdot x^3 = x^{5+3} = x^8$

109. Simplifying: $\frac{x^3}{x^2} = x^{3-2} = x$

111. Simplifying: $\frac{y^3}{y^5} = y^{3-5} = y^{-2} = \frac{1}{y^2}$

113. Writing in expanded form: $3.4 \times 10^2 = 340$

4.3 Operations with Monomials

1.   Multiplying the monomials: $(3x^4)(4x^3) = 12x^{4+3} = 12x^7$

3.   Multiplying the monomials: $(-2y^4)(8y^7) = -16y^{4+7} = -16y^{11}$

5.   Multiplying the monomials: $(8x)(4x) = 32x^{1+1} = 32x^2$

7.   Multiplying the monomials: $(10a^3)(10a)(2a^2) = 200a^{3+1+2} = 200a^6$

9.   Multiplying the monomials: $(6ab^2)(-4a^2b) = -24a^{1+2}b^{2+1} = -24a^3b^3$

11. Multiplying the monomials: $(4x^2y)(3x^3y^3)(2xy^4) = 24x^{2+3+1}y^{1+3+4} = 24x^6y^8$

13. Dividing the monomials: $\dfrac{15x^3}{5x^2} = \dfrac{15}{5} \cdot \dfrac{x^3}{x^2} = 3x$

15. Dividing the monomials: $\dfrac{18y^9}{3y^{12}} = \dfrac{18}{3} \cdot \dfrac{y^9}{y^{12}} = 6 \cdot \dfrac{1}{y^3} = \dfrac{6}{y^3}$

17. Dividing the monomials: $\dfrac{32a^3}{64a^4} = \dfrac{32}{64} \cdot \dfrac{a^3}{a^4} = \dfrac{1}{2} \cdot \dfrac{1}{a} = \dfrac{1}{2a}$

19. Dividing the monomials: $\dfrac{21a^2b^3}{-7ab^5} = \dfrac{21}{-7} \cdot \dfrac{a^2}{a} \cdot \dfrac{b^3}{b^5} = -3 \cdot a \cdot \dfrac{1}{b^2} = -\dfrac{3a}{b^2}$

21. Dividing the monomials: $\dfrac{3x^3y^2z}{27xy^2z^3} = \dfrac{3}{27} \cdot \dfrac{x^3}{x} \cdot \dfrac{y^2}{y^2} \cdot \dfrac{z}{z^3} = \dfrac{1}{9} \cdot x^2 \cdot \dfrac{1}{z^2} = \dfrac{x^2}{9z^2}$

23. Completing the table:

a	b	ab	$\dfrac{a}{b}$	$\dfrac{b}{a}$
10	$5x$	$50x$	$\dfrac{2}{x}$	$\dfrac{x}{2}$
$20x^3$	$6x^2$	$120x^5$	$\dfrac{10x}{3}$	$\dfrac{3}{10x}$
$25x^5$	$5x^4$	$125x^9$	$5x$	$\dfrac{1}{5x}$
$3x^{-2}$	$3x^2$	9	$\dfrac{1}{x^4}$	x^4
$-2y^4$	$8y^7$	$-16y^{11}$	$-\dfrac{1}{4y^3}$	$-4y^3$

25. Finding the product: $(3 \times 10^3)(2 \times 10^5) = 6 \times 10^8$

27. Finding the product: $(3.5 \times 10^4)(5 \times 10^{-6}) = 17.5 \times 10^{-2} = 1.75 \times 10^{-1}$

29. Finding the product: $(5.5 \times 10^{-3})(2.2 \times 10^{-4}) = 12.1 \times 10^{-7} = 1.21 \times 10^{-6}$

31. Finding the quotient: $\dfrac{8.4 \times 10^5}{2 \times 10^2} = 4.2 \times 10^3$

33. Finding the quotient: $\dfrac{6 \times 10^8}{2 \times 10^{-2}} = 3 \times 10^{10}$

35. Finding the quotient: $\dfrac{2.5 \times 10^{-6}}{5 \times 10^{-4}} = 0.5 \times 10^{-2} = 5 \times 10^{-3}$

37. Combining the monomials: $3x^2 + 5x^2 = (3+5)x^2 = 8x^2$

39. Combining the monomials: $8x^5 - 19x^5 = (8-19)x^5 = -11x^5$

41. Combining the monomials: $2a + a - 3a = (2+1-3)a = 0a = 0$

43. Combining the monomials: $10x^3 - 8x^3 + 2x^3 = (10-8+2)x^3 = 4x^3$

45. Combining the monomials: $20ab^2 - 19ab^2 + 30ab^2 = (20-19+30)ab^2 = 31ab^2$

47. Completing the table:

a	b	ab	$a+b$
$5x$	$3x$	$15x^2$	$8x$
$4x^2$	$2x^2$	$8x^4$	$6x^2$
$3x^3$	$6x^3$	$18x^6$	$9x^3$
$2x^4$	$-3x^4$	$-6x^8$	$-x^4$
x^5	$7x^5$	$7x^{10}$	$8x^5$

49. Simplifying the expression: $\dfrac{\left(3x^2\right)\left(8x^5\right)}{6x^4} = \dfrac{24x^7}{6x^4} = \dfrac{24}{6} \cdot \dfrac{x^7}{x^4} = 4x^3$

51. Simplifying the expression: $\dfrac{\left(9a^2b\right)\left(2a^3b^4\right)}{18a^5b^7} = \dfrac{18a^5b^5}{18a^5b^7} = \dfrac{18}{18} \cdot \dfrac{a^5}{a^5} \cdot \dfrac{b^5}{b^7} = 1 \cdot \dfrac{1}{b^2} = \dfrac{1}{b^2}$

53. Simplifying the expression: $\dfrac{\left(4x^3y^2\right)\left(9x^4y^{10}\right)}{\left(3x^5y\right)\left(2x^6y\right)} = \dfrac{36x^7y^{12}}{6x^{11}y^2} = \dfrac{36}{6} \cdot \dfrac{x^7}{x^{11}} \cdot \dfrac{y^{12}}{y^2} = 6 \cdot \dfrac{1}{x^4} \cdot y^{10} = \dfrac{6y^{10}}{x^4}$

55. Applying the distributive property: $xy\left(x + \dfrac{1}{y}\right) = xy \cdot x + xy \cdot \dfrac{1}{y} = x^2y + x$

57. Applying the distributive property: $xy\left(\dfrac{1}{y} + \dfrac{1}{x}\right) = xy \cdot \dfrac{1}{y} + xy \cdot \dfrac{1}{x} = x + y$

59. Applying the distributive property: $x^2\left(1 - \dfrac{4}{x^2}\right) = x^2 \cdot 1 - x^2 \cdot \dfrac{4}{x^2} = x^2 - 4$

61. Applying the distributive property: $x^2\left(1 - \dfrac{1}{x} - \dfrac{6}{x^2}\right) = x^2 \cdot 1 - x^2 \cdot \dfrac{1}{x} - x^2 \cdot \dfrac{6}{x^2} = x^2 - x - 6$

63. Applying the distributive property: $x^2\left(1 - \dfrac{5}{x}\right) = x^2 \cdot 1 - x^2 \cdot \dfrac{5}{x} = x^2 - 5x$

65. Applying the distributive property: $x^2\left(1 - \dfrac{8}{x}\right) = x^2 \cdot 1 - x^2 \cdot \dfrac{8}{x} = x^2 - 8x$

67. Simplifying the expression: $\dfrac{\left(6\times10^{8}\right)\left(3\times10^{5}\right)}{9\times10^{7}}=\dfrac{18\times10^{13}}{9\times10^{7}}=2\times10^{6}$

69. Simplifying the expression: $\dfrac{\left(5\times10^{3}\right)\left(4\times10^{-5}\right)}{2\times10^{-2}}=\dfrac{20\times10^{-2}}{2\times10^{-2}}=10=1\times10^{1}$

71. Simplifying the expression: $\dfrac{\left(2.8\times10^{-7}\right)\left(3.6\times10^{4}\right)}{2.4\times10^{3}}=\dfrac{10.08\times10^{-3}}{2.4\times10^{3}}=4.2\times10^{-6}$

73. Simplifying the expression: $\dfrac{18x^{4}}{3x}+\dfrac{21x^{7}}{7x^{4}}=6x^{3}+3x^{3}=9x^{3}$

75. Simplifying the expression: $\dfrac{45a^{6}}{9a^{4}}-\dfrac{50a^{8}}{2a^{6}}=5a^{2}-25a^{2}=-20a^{2}$

77. Simplifying the expression: $\dfrac{6x^{7}y^{4}}{3x^{2}y^{2}}+\dfrac{8x^{5}y^{8}}{2y^{6}}=2x^{5}y^{2}+4x^{5}y^{2}=6x^{5}y^{2}$

79. Simplifying: $3-8=-5$ 81. Simplifying: $-1+7=6$

83. Simplifying: $3(5)^{2}+1=3(25)+1=75+1=76$

85. Simplifying: $2x^{2}+4x^{2}=6x^{2}$ 87. Simplifying: $-5x+7x=2x$

89. Simplifying: $-(2x+9)=-2x-9$

91. Substituting $x=4$: $2x+3=2(4)+3=8+3=11$

4.4 Addition and Subtraction of Polynomials

1. This is a trinomial of degree 3.   3. This is a trinomial of degree 3.
5. This is a binomial of degree 1.   7. This is a binomial of degree 2.
9. This is a monomial of degree 2.   11. This is a monomial of degree 0.
13. Combining the polynomials:

$$\left(2x^{2}+3x+4\right)+\left(3x^{2}+2x+5\right)=\left(2x^{2}+3x^{2}\right)+\left(3x+2x\right)+\left(4+5\right)=5x^{2}+5x+9$$

15. Combining the polynomials:

$$\left(3a^{2}-4a+1\right)+\left(2a^{2}-5a+6\right)=\left(3a^{2}+2a^{2}\right)+\left(-4a-5a\right)+\left(1+6\right)=5a^{2}-9a+7$$

17. Combining the polynomials: $x^{2}+4x+2x+8=x^{2}+\left(4x+2x\right)+8=x^{2}+6x+8$

19. Combining the polynomials: $6x^{2}-3x-10x+5=6x^{2}+\left(-3x-10x\right)+5=6x^{2}-13x+5$

21. Combining the polynomials: $x^{2}-3x+3x-9=x^{2}+\left(-3x+3x\right)-9=x^{2}-9$

23. Combining the polynomials: $3y^{2}-5y-6y+10=3y^{2}+\left(-5y-6y\right)+10=3y^{2}-11y+10$

25. Combining the polynomials:

$$\left(6x^{3}-4x^{2}+2x\right)+\left(9x^{2}-6x+3\right)=6x^{3}+\left(-4x^{2}+9x^{2}\right)+\left(2x-6x\right)+3=6x^{3}+5x^{2}-4x+3$$

27. Combining the polynomials:

$$\left(\dfrac{2}{3}x^{2}-\dfrac{1}{5}x-\dfrac{3}{4}\right)+\left(\dfrac{4}{3}x^{2}-\dfrac{4}{5}x+\dfrac{7}{4}\right)=\left(\dfrac{2}{3}x^{2}+\dfrac{4}{3}x^{2}\right)+\left(-\dfrac{1}{5}x-\dfrac{4}{5}x\right)+\left(-\dfrac{3}{4}+\dfrac{7}{4}\right)$$
$$=2x^{2}-x+1$$

29. Combining the polynomials: $\left(a^{2}-a-1\right)-\left(-a^{2}+a+1\right)=a^{2}-a-1+a^{2}-a-1=2a^{2}-2a-2$

31. Combining the polynomials:

$$\left(\frac{5}{9}x^3 + \frac{1}{3}x^2 - 2x + 1\right) - \left(\frac{2}{3}x^3 + x^2 + \frac{1}{2}x - \frac{3}{4}\right) = \frac{5}{9}x^3 + \frac{1}{3}x^2 - 2x + 1 - \frac{2}{3}x^3 - x^2 - \frac{1}{2}x + \frac{3}{4}$$

$$= -\frac{1}{9}x^3 - \frac{2}{3}x^2 - \frac{5}{2}x + \frac{7}{4}$$

33. Combining the polynomials:

$$\left(4y^2 - 3y + 2\right) + \left(5y^2 + 12y - 4\right) - \left(13y^2 - 6y + 20\right)$$

$$= 4y^2 - 3y + 2 + 5y^2 + 12y - 4 - 13y^2 + 6y - 20$$

$$= \left(4y^2 + 5y^2 - 13y^2\right) + \left(-3y + 12y + 6y\right) + \left(2 - 4 - 20\right)$$

$$= -4y^2 + 15y - 22$$

35. Performing the subtraction:

$$\left(11x^2 - 10x + 13\right) - \left(10x^2 + 23x - 50\right) = 11x^2 - 10x + 13 - 10x^2 - 23x + 50$$

$$= \left(11x^2 - 10x^2\right) + \left(-10x - 23x\right) + \left(13 + 50\right)$$

$$= x^2 - 33x + 63$$

37. Performing the subtraction:

$$\left(11y^2 + 11y + 11\right) - \left(3y^2 + 7y - 15\right) = 11y^2 + 11y + 11 - 3y^2 - 7y + 15$$

$$= \left(11y^2 - 3y^2\right) + \left(11y - 7y\right) + \left(11 + 15\right)$$

$$= 8y^2 + 4y + 26$$

39. Performing the addition:

$$\left(25x^2 - 50x + 75\right) + \left(50x^2 - 100x - 150\right) = \left(25x^2 + 50x^2\right) + \left(-50x - 100x\right) + \left(75 - 150\right)$$

$$= 75x^2 - 150x - 75$$

41. Performing the operations:

$$\left(3x - 2\right) + \left(11x + 5\right) - \left(2x + 1\right) = 3x - 2 + 11x + 5 - 2x - 1$$

$$= \left(3x + 11x - 2x\right) + \left(-2 + 5 - 1\right)$$

$$= 12x + 2$$

43. Evaluating when $x = 3$: $x^2 - 2x + 1 = (3)^2 - 2(3) + 1 = 9 - 6 + 1 = 4$

45. Finding the volume of the cylinder and sphere:

$$V_{\text{cylinder}} = \pi\left(3^2\right)(6) = 54\pi \qquad\qquad V_{\text{sphere}} = \frac{4}{3}\pi\left(3^3\right) = 36\pi$$

Subtracting to find the amount of space to pack: $V = 54\pi - 36\pi = 18\pi \approx 56.52$ inches^3

47. Simplifying: $(-5)(-1) = 5$ **49.** Simplifying: $(-1)(6) = -6$

51. Simplifying: $(5x)(-4x) = -20x^2$ **53.** Simplifying: $3x(-7) = -21x$

55. Simplifying: $5x + (-3x) = 2x$ **57.** Multiplying: $3(2x - 6) = 6x - 18$

4.5 Multiplication with Polynomials

1. Using the distributive property: $2x(3x+1)=2x(3x)+2x(1)=6x^2+2x$
3. Using the distributive property:
$$2x^2\left(3x^2-2x+1\right)=2x^2\left(3x^2\right)-2x^2\left(2x\right)+2x^2(1)=6x^4-4x^3+2x^2$$
5. Using the distributive property:
$$2ab\left(a^2-ab+1\right)=2ab\left(a^2\right)-2ab(ab)+2ab(1)=2a^3b-2a^2b^2+2ab$$
7. Using the distributive property: $y^2\left(3y^2+9y+12\right)=y^2\left(3y^2\right)+y^2\left(9y\right)+y^2\left(12\right)=3y^4+9y^3+12y^2$
9. Using the distributive property:
$$4x^2y\left(2x^3y+3x^2y^2+8y^3\right)=4x^2y\left(2x^3y\right)+4x^2y\left(3x^2y^2\right)+4x^2y\left(8y^3\right)$$
$$=8x^5y^2+12x^4y^3+32x^2y^4$$
11. Multiplying using the FOIL method: $(x+3)(x+4)=x^2+3x+4x+12=x^2+7x+12$
13. Multiplying using the FOIL method: $(x+6)(x+1)=x^2+6x+1x+6=x^2+7x+6$
15. Multiplying using the FOIL method: $\left(x+\dfrac{1}{2}\right)\left(x+\dfrac{3}{2}\right)=x^2+\dfrac{1}{2}x+\dfrac{3}{2}x+\dfrac{3}{4}=x^2+2x+\dfrac{3}{4}$
17. Multiplying using the FOIL method: $(a+5)(a-3)=a^2+5a-3a-15=a^2+2a-15$
19. Multiplying using the FOIL method: $(x-a)(y+b)=xy-ay+bx-ab$
21. Multiplying using the FOIL method: $(x+6)(x-6)=x^2+6x-6x-36=x^2-36$
23. Multiplying using the FOIL method: $\left(y+\dfrac{5}{6}\right)\left(y-\dfrac{5}{6}\right)=y^2+\dfrac{5}{6}y-\dfrac{5}{6}y-\dfrac{25}{36}=y^2-\dfrac{25}{36}$
25. Multiplying using the FOIL method: $(2x-3)(x-4)=2x^2-3x-8x+12=2x^2-11x+12$
27. Multiplying using the FOIL method: $(a+2)(2a-1)=2a^2+4a-a-2=2a^2+3a-2$
29. Multiplying using the FOIL method: $(2x-5)(3x-2)=6x^2-15x-4x+10=6x^2-19x+10$
31. Multiplying using the FOIL method: $(2x+3)(a+4)=2ax+3a+8x+12$
33. Multiplying using the FOIL method: $(5x-4)(5x+4)=25x^2-20x+20x-16=25x^2-16$
35. Multiplying using the FOIL method: $\left(2x-\dfrac{1}{2}\right)\left(x+\dfrac{3}{2}\right)=2x^2-\dfrac{1}{2}x+3x-\dfrac{3}{4}=2x^2+\dfrac{5}{2}x-\dfrac{3}{4}$
37. Multiplying using the FOIL method: $(1-2a)(3-4a)=3-6a-4a+8a^2=3-10a+8a^2$
39. The product is $(x+2)(x+3)=x^2+5x+6$:   41. The product is $(x+1)(2x+2)=2x^2+4x+2$:

	x	3
x	x^2	$3x$
2	$2x$	6

	x	x	2
x	x^2	x^2	$2x$
1	x	x	2

43. Multiplying using the column method:

$$a^2 - 3a + 2$$
$$\underline{ a - 3}$$
$$a^3 - 3a^2 + 2a$$
$$\underline{ -3a^2 + 9a - 6}$$
$$a^3 - 6a^2 + 11a - 6$$

45. Multiplying using the column method:

$$x^2 - 2x + 4$$
$$\underline{ x + 2}$$
$$x^3 - 2x^2 + 4x$$
$$\underline{ 2x^2 - 4x + 8}$$
$$x^3 + 8$$

47. Multiplying using the column method:

$$x^2 + 8x + 9$$
$$\underline{ 2x + 1}$$
$$2x^3 + 16x^2 + 18x$$
$$\underline{ x^2 + 8x + 9}$$
$$2x^3 + 17x^2 + 26x + 9$$

49. Multiplying using the column method:

$$5x^2 + 2x + 1$$
$$\underline{ x^2 - 3x + 5}$$
$$5x^4 + 2x^3 + x^2$$
$$ -15x^3 - 6x^2 - 3x$$
$$\underline{ 25x^2 + 10x + 5}$$
$$5x^4 - 13x^3 + 20x^2 + 7x + 5$$

51. Multiplying using the FOIL method: $(x^2 + 3)(2x^2 - 5) = 2x^4 - 5x^2 + 6x^2 - 15 = 2x^4 + x^2 - 15$

53. Multiplying using the FOIL method: $(3a^4 + 2)(2a^2 + 5) = 6a^6 + 15a^4 + 4a^2 + 10$

55. First multiply two polynomials using the FOIL method:
$$(x + 3)(x + 4) = x^2 + 3x + 4x + 12 = x^2 + 7x + 12$$

Now using the column method:

$$x^2 + 7x + 12$$
$$\underline{ x + 5}$$
$$x^3 + 7x^2 + 12x$$
$$\underline{ 5x^2 + 35x + 60}$$
$$x^3 + 12x^2 + 47x + 60$$

57. Simplifying: $(x - 3)(x - 2) + 2 = x^2 - 3x - 2x + 6 + 2 = x^2 - 5x + 8$

59. Simplifying: $(2x - 3)(4x + 3) + 4 = 8x^2 + 6x - 12x - 9 + 4 = 8x^2 - 6x - 5$

61. Simplifying: $(x + 4)(x - 5) + (-5)(2) = x^2 - 5x + 4x - 20 - 10 = x^2 - x - 30$

63. Simplifying: $2(x - 3) + x(x + 2) = 2x - 6 + x^2 + 2x = x^2 + 4x - 6$

65. Simplifying: $3x(x + 1) - 2x(x - 5) = 3x^2 + 3x - 2x^2 + 10x = x^2 + 13x$

67. Simplifying: $x(x + 2) - 3 = x^2 + 2x - 3$ **69.** Simplifying: $a(a - 3) + 6 = a^2 - 3a + 6$

71. Let x represent the width and $2x + 5$ represent the length. The area is given by:
$$A = x(2x + 5) = 2x^2 + 5x$$

73. Let x and $x + 1$ represent the width and length, respectively. The area is given by:
$$A = x(x + 1) = x^2 + x$$

75. Simplifying: $13 \cdot 13 = 169$ **77.** Simplifying: $2(x)(-5) = -10x$

79. Simplifying: $6x + (-6x) = 0$

81. Simplifying: $(2x)(-3)+(2x)(3)=-6x+6x=0$
83. Multiplying: $-4(3x-4)=-12x+16$
85. Multiplying: $(x-1)(x+2)=x^2+2x-x-2=x^2+x-2$
87. Multiplying: $(x+3)(x+3)=x^2+3x+3x+9=x^2+6x+9$

4.6 Binomial Squares and Other Special Products

1. Multiplying using the FOIL method: $(x-2)^2=(x-2)(x-2)=x^2-2x-2x+4=x^2-4x+4$
3. Multiplying using the FOIL method: $(a+3)^2=(a+3)(a+3)=a^2+3a+3a+9=a^2+6a+9$
5. Multiplying using the FOIL method: $(x-5)^2=(x-5)(x-5)=x^2-5x-5x+25=x^2-10x+25$
7. Multiplying using the FOIL method:
$$\left(a-\frac{1}{2}\right)^2=\left(a-\frac{1}{2}\right)\left(a-\frac{1}{2}\right)=a^2-\frac{1}{2}a-\frac{1}{2}a+\frac{1}{4}=a^2-a+\frac{1}{4}$$
9. Multiplying using the FOIL method:
$$(x+10)^2=(x+10)(x+10)=x^2+10x+10x+100=x^2+20x+100$$
11. Multiplying using the square of binomial formula:
$$(a+0.8)^2=a^2+2(a)(0.8)+(0.8)^2=a^2+1.6a+0.64$$
13. Multiplying using the square of binomial formula:
$$(2x-1)^2=(2x)^2-2(2x)(1)+(1)^2=4x^2-4x+1$$
15. Multiplying using the square of binomial formula:
$$(4a+5)^2=(4a)^2+2(4a)(5)+(5)^2=16a^2+40a+25$$
17. Multiplying using the square of binomial formula:
$$(3x-2)^2=(3x)^2-2(3x)(2)+(2)^2=9x^2-12x+4$$
19. Multiplying using the square of binomial formula:
$$(3a+5b)^2=(3a)^2+2(3a)(5b)+(5b)^2=9a^2+30ab+25b^2$$
21. Multiplying using the square of binomial formula:
$$(4x-5y)^2=(4x)^2-2(4x)(5y)+(5y)^2=16x^2-40xy+25y^2$$
23. Multiplying using the square of binomial formula:
$$(7m+2n)^2=(7m)^2+2(7m)(2n)+(2n)^2=49m^2+28mn+4n^2$$
25. Multiplying using the square of binomial formula:
$$(6x-10y)^2=(6x)^2-2(6x)(10y)+(10y)^2=36x^2-120xy+100y^2$$
27. Multiplying using the square of binomial formula:
$$\left(x^2+5\right)^2=\left(x^2\right)^2+2\left(x^2\right)(5)+(5)^2=x^4+10x^2+25$$
29. Multiplying using the square of binomial formula:
$$\left(a^2+1\right)^2=\left(a^2\right)^2+2\left(a^2\right)(1)+(1)^2=a^4+2a^2+1$$

31. Completing the table:

x	$(x+3)^2$	x^2+9	x^2+6x+9
1	16	10	16
2	25	13	25
3	36	18	36
4	49	25	49

33. Completing the table:

a	1	3	3	4
b	1	5	4	5
$(a+b)^2$	4	64	49	81
a^2+b^2	2	34	25	41
a^2+ab+b^2	3	49	37	61
$a^2+2ab+b^2$	4	64	49	81

35. Multiplying using the FOIL method: $(a+5)(a-5) = a^2 + 5a - 5a - 25 = a^2 - 25$

37. Multiplying using the FOIL method: $(y-1)(y+1) = y^2 - y + y - 1 = y^2 - 1$

39. Multiplying using the difference of squares formula: $(9+x)(9-x) = (9)^2 - (x)^2 = 81 - x^2$

41. Multiplying using the difference of squares formula: $(2x+5)(2x-5) = (2x)^2 - (5)^2 = 4x^2 - 25$

43. Multiplying using the difference of squares formula:

$$\left(4x+\frac{1}{3}\right)\left(4x-\frac{1}{3}\right) = (4x)^2 - \left(\frac{1}{3}\right)^2 = 16x^2 - \frac{1}{9}$$

45. Multiplying using the difference of squares formula:

$$(2a+7)(2a-7) = (2a)^2 - (7)^2 = 4a^2 - 49$$

47. Multiplying using the difference of squares formula: $(6-7x)(6+7x) = (6)^2 - (7x)^2 = 36 - 49x^2$

49. Multiplying using the difference of squares formula: $(x^2+3)(x^2-3) = (x^2)^2 - (3)^2 = x^4 - 9$

51. Multiplying using the difference of squares formula: $(a^2+4)(a^2-4) = (a^2)^2 - (4)^2 = a^4 - 16$

53. Multiplying using the difference of squares formula:

$$(5y^4-8)(5y^4+8) = (5y^4)^2 - (8)^2 = 25y^8 - 64$$

55. Multiplying and simplifying: $(x+3)(x-3)+(x+5)(x-5) = (x^2-9)+(x^2-25) = 2x^2 - 34$

57. Multiplying and simplifying:

$$(2x+3)^2 - (4x-1)^2 = (4x^2+12x+9) - (16x^2-8x+1)$$
$$= 4x^2 + 12x + 9 - 16x^2 + 8x - 1$$
$$= -12x^2 + 20x + 8$$

59. Multiplying and simplifying:

$$(a+1)^2 - (a+2)^2 + (a+3)^2 = (a^2+2a+1) - (a^2+4a+4) + (a^2+6a+9)$$
$$= a^2 + 2a + 1 - a^2 - 4a - 4 + a^2 + 6a + 9$$
$$= a^2 + 4a + 6$$

61. Multiplying and simplifying:

$$(2x+3)^3 = (2x+3)(2x+3)^2$$
$$= (2x+3)(4x^2+12x+9)$$
$$= 8x^3 + 24x^2 + 18x + 12x^2 + 36x + 27$$
$$= 8x^3 + 36x^2 + 54x + 27$$

63. Finding the product: $49(51) = (50-1)(50+1) = (50)^2 - (1)^2 = 2{,}500 - 1 = 2{,}499$

65. Evaluating when $x = 2$:

$$(x+3)^2 = (2+3)^2 = (5)^2 = 25$$

$$x^2 + 6x + 9 = (2)^2 + 6(2) + 9 = 4 + 12 + 9 = 25$$

67. Let x and $x+1$ represent the two integers. The expression can be written as:

$$(x)^2 + (x+1)^2 = x^2 + (x^2 + 2x + 1) = 2x^2 + 2x + 1$$

69. Let x, $x+1$, and $x+2$ represent the three integers. The expression can be written as:

$$(x)^2 + (x+1)^2 + (x+2)^2 = x^2 + (x^2 + 2x + 1) + (x^2 + 4x + 4) = 3x^2 + 6x + 5$$

71. Verifying the areas: $(a+b)^2 = a^2 + ab + ab + b^2 = a^2 + 2ab + b^2$

73. Simplifying: $\dfrac{10x^3}{5x} = 2x^{3-1} = 2x^2$

75. Simplifying: $\dfrac{3x^2}{3} = x^2$

77. Simplifying: $\dfrac{9x^2}{3x} = 3x^{2-1} = 3x$

79. Simplifying: $\dfrac{24x^3y^2}{8x^2y} = 3x^{3-2}y^{2-1} = 3xy$

4.7 Dividing a Polynomial by a Monomial

1. Performing the division: $\dfrac{5x^2 - 10x}{5x} = \dfrac{5x^2}{5x} - \dfrac{10x}{5x} = x - 2$

3. Performing the division: $\dfrac{15x - 10x^3}{5x} = \dfrac{15x}{5x} - \dfrac{10x^3}{5x} = 3 - 2x^2$

5. Performing the division: $\dfrac{25x^2y - 10xy}{5x} = \dfrac{25x^2y}{5x} - \dfrac{10xy}{5x} = 5xy - 2y$

7. Performing the division: $\dfrac{35x^5 - 30x^4 + 25x^3}{5x} = \dfrac{35x^5}{5x} - \dfrac{30x^4}{5x} + \dfrac{25x^3}{5x} = 7x^4 - 6x^3 + 5x^2$

9. Performing the division: $\dfrac{50x^5 - 25x^3 + 5x}{5x} = \dfrac{50x^5}{5x} - \dfrac{25x^3}{5x} + \dfrac{5x}{5x} = 10x^4 - 5x^2 + 1$

11. Performing the division: $\dfrac{8a^2 - 4a}{-2a} = \dfrac{8a^2}{-2a} + \dfrac{-4a}{-2a} = -4a + 2$

13. Performing the division: $\dfrac{16a^5 + 24a^4}{-2a} = \dfrac{16a^5}{-2a} + \dfrac{24a^4}{-2a} = -8a^4 - 12a^3$

15. Performing the division: $\dfrac{8ab + 10a^2}{-2a} = \dfrac{8ab}{-2a} + \dfrac{10a^2}{-2a} = -4b - 5a$

17. Performing the division: $\dfrac{12a^3b - 6a^2b^2 + 14ab^3}{-2a} = \dfrac{12a^3b}{-2a} + \dfrac{-6a^2b^2}{-2a} + \dfrac{14ab^3}{-2a} = -6a^2b + 3ab^2 - 7b^3$

19. Performing the division: $\dfrac{a^2 + 2ab + b^2}{-2a} = \dfrac{a^2}{-2a} + \dfrac{2ab}{-2a} + \dfrac{b^2}{-2a} = -\dfrac{a}{2} - b - \dfrac{b^2}{2a}$

21. Performing the division: $\dfrac{6x + 8y}{2} = \dfrac{6x}{2} + \dfrac{8y}{2} = 3x + 4y$

23.   Performing the division: $\dfrac{7y-21}{-7} = \dfrac{7y}{-7} + \dfrac{-21}{-7} = -y+3$

25.   Performing the division: $\dfrac{10xy-8x}{2x} = \dfrac{10xy}{2x} - \dfrac{8x}{2x} = 5y-4$

27.   Performing the division: $\dfrac{x^2y-x^3y^2}{x} = \dfrac{x^2y}{x} - \dfrac{x^3y^2}{x} = xy - x^2y^2$

29.   Performing the division: $\dfrac{x^2y-x^3y^2}{-x^2y} = \dfrac{x^2y}{-x^2y} + \dfrac{-x^3y^2}{-x^2y} = -1+xy$

31.   Performing the division: $\dfrac{a^2b^2-ab^2}{-ab^2} = \dfrac{a^2b^2}{-ab^2} + \dfrac{-ab^2}{-ab^2} = -a+1$

33.   Performing the division: $\dfrac{x^3-3x^2y+xy^2}{x} = \dfrac{x^3}{x} - \dfrac{3x^2y}{x} + \dfrac{xy^2}{x} = x^2 - 3xy + y^2$

35.   Performing the division: $\dfrac{10a^2-15a^2b+25a^2b^2}{5a^2} = \dfrac{10a^2}{5a^2} - \dfrac{15a^2b}{5a^2} + \dfrac{25a^2b^2}{5a^2} = 2-3b+5b^2$

37.   Performing the division: $\dfrac{26x^2y^2-13xy}{-13xy} = \dfrac{26x^2y^2}{-13xy} + \dfrac{-13xy}{-13xy} = -2xy+1$

39.   Performing the division: $\dfrac{4x^2y^2-2xy}{4xy} = \dfrac{4x^2y^2}{4xy} - \dfrac{2xy}{4xy} = xy - \dfrac{1}{2}$

41.   Performing the division: $\dfrac{5a^2x-10ax^2+15a^2x^2}{20a^2x^2} = \dfrac{5a^2x}{20a^2x^2} - \dfrac{10ax^2}{20a^2x^2} + \dfrac{15a^2x^2}{20a^2x^2} = \dfrac{1}{4x} - \dfrac{1}{2a} + \dfrac{3}{4}$

43.   Performing the division: $\dfrac{16x^5+8x^2+12x}{12x^3} = \dfrac{16x^5}{12x^3} + \dfrac{8x^2}{12x^3} + \dfrac{12x}{12x^3} = \dfrac{4x^2}{3} + \dfrac{2}{3x} + \dfrac{1}{x^2}$

45.   Performing the division: $\dfrac{9a^{5m}-27a^{3m}}{3a^{2m}} = \dfrac{9a^{5m}}{3a^{2m}} - \dfrac{27a^{3m}}{3a^{2m}} = 3a^{5m-2m} - 9a^{3m-2m} = 3a^{3m} - 9a^{m}$

47.   Performing the division:

$$\dfrac{10x^{5m}-25x^{3m}+35x^m}{5x^m} = \dfrac{10x^{5m}}{5x^m} - \dfrac{25x^{3m}}{5x^m} + \dfrac{35x^m}{5x^m}$$

$$= 2x^{5m-m} - 5x^{3m-m} + 7x^{m-m}$$

$$= 2x^{4m} - 5x^{2m} + 7$$

49.   Simplifying and then dividing:

$$\dfrac{2x^3(3x+2)-3x^2(2x-4)}{2x^2} = \dfrac{6x^4+4x^3-6x^3+12x^2}{2x^2}$$

$$= \dfrac{6x^4-2x^3+12x^2}{2x^2}$$

$$= \dfrac{6x^4}{2x^2} - \dfrac{2x^3}{2x^2} + \dfrac{12x^2}{2x^2}$$

$$= 3x^2 - x + 6$$

51. Simplifying and then dividing:

$$\frac{(x+2)^2-(x-2)^2}{2x}=\frac{\left(x^2+4x+4\right)-\left(x^2-4x+4\right)}{2x}=\frac{x^2+4x+4-x^2+4x-4}{2x}=\frac{8x}{2x}=4$$

53. Simplifying and then dividing:

$$\frac{(x+5)^2+(x+5)(x-5)}{2x}=\frac{\left(x^2+10x+25\right)+\left(x^2-25\right)}{2x}=\frac{2x^2+10x}{2x}=\frac{2x^2}{2x}+\frac{10x}{2x}=x+5$$

55. Evaluating each expression when $x=2$:

$$\frac{10x+15}{5}=\frac{10(2)+15}{5}=\frac{20+15}{5}=\frac{35}{5}=7 \qquad\qquad 2x+3=2(2)+3=4+3=7$$

57. Evaluating each expression when $x=10$:

$$\frac{3x+8}{2}=\frac{3(10)+8}{2}=\frac{30+8}{2}=\frac{38}{2}=19 \qquad\qquad 3x+4=3(10)+4=30+4=34$$

Thus $\dfrac{3x+8}{2}\neq 3x+4$.

59. Dividing:

$$
\begin{array}{r}
146 \\
27\overline{)3962} \\
\underline{27} \\
126 \\
\underline{108} \\
182 \\
\underline{162} \\
20
\end{array}
$$

The quotient is $146\dfrac{20}{27}$.

61. Dividing: $\dfrac{2x^2+5x}{x}=\dfrac{2x^2}{x}+\dfrac{5x}{x}=2x+5$ **63.** Multiplying: $(x-3)x=x^2-3x$

65. Multiplying: $2x^2(x-5)=2x^3-10x^2$

67. Subtracting: $\left(x^2-5x\right)-\left(x^2-3x\right)=x^2-5x-x^2+3x=-2x$

69. Subtracting: $(-2x+8)-(-2x+6)=-2x+8+2x-6=2$

4.8 Dividing a Polynomial by a Polynomial

1. Using long division:

$$\begin{array}{r} x-2 \\ x-3\overline{\smash{\big)}\,x^2-5x+6} \\ \underline{x^2-3x} \\ -2x+6 \\ \underline{-2x+6} \\ 0 \end{array}$$

The quotient is $x-2$.

3. Using long division:

$$\begin{array}{r} a+4 \\ a+5\overline{\smash{\big)}\,a^2+9a+20} \\ \underline{a^2+5a} \\ 4a+20 \\ \underline{4a+20} \\ 0 \end{array}$$

The quotient is $a+4$.

5. Using long division:

$$\begin{array}{r} x-3 \\ x-3\overline{\smash{\big)}\,x^2-6x+9} \\ \underline{x^2-3x} \\ -3x+9 \\ \underline{-3x+9} \\ 0 \end{array}$$

The quotient is $x-3$.

7. Using long division:

$$\begin{array}{r} x+3 \\ 2x-1\overline{\smash{\big)}\,2x^2+5x-3} \\ \underline{2x^2-x} \\ 6x-3 \\ \underline{6x-3} \\ 0 \end{array}$$

The quotient is $x+3$.

9. Using long division:

$$\begin{array}{r} a-5 \\ 2a+1\overline{\smash{\big)}\,2a^2-9a-5} \\ \underline{2a^2+a} \\ -10a-5 \\ \underline{-10a-5} \\ 0 \end{array}$$

The quotient is $a-5$.

11. Using long division:

$$\begin{array}{r} x+2 \\ x+3\overline{\smash{\big)}\,x^2+5x+8} \\ \underline{x^2+3x} \\ 2x+8 \\ \underline{2x+6} \\ 2 \end{array}$$

The quotient is $x+2+\dfrac{2}{x+3}$.

13. Using long division:

$$\begin{array}{r} a-2 \\ a+5\overline{\smash{\big)}\,a^2+3a+2} \\ \underline{a^2+5a} \\ -2a+2 \\ \underline{-2a-10} \\ 12 \end{array}$$

The quotient is $a-2+\dfrac{12}{a+5}$.

15. Using long division:

$$\begin{array}{r} x+4 \\ x-2\overline{\smash{\big)}\,x^2+2x+1} \\ \underline{x^2-2x} \\ 4x+1 \\ \underline{4x-8} \\ 9 \end{array}$$

The quotient is $x+4+\dfrac{9}{x-2}$.

17. Using long division:

$$x+1\overline{)x^2+5x-6}$$

quotient: $x+4$

$$\underline{x^2+\ x}$$
$$4x-6$$
$$\underline{4x+4}$$
$$-10$$

The quotient is $x+4+\dfrac{-10}{x+1}$.

19. Using long division:

$$a+2\overline{)a^2+3a+1}$$

quotient: $a+1$

$$\underline{a^2+2a}$$
$$a+1$$
$$\underline{a+2}$$
$$-1$$

The quotient is $a+1+\dfrac{-1}{a+2}$.

21. Using long division:

$$2x+4\overline{)2x^2-2x+5}$$

quotient: $x-3$

$$\underline{2x^2+4x}$$
$$-6x+5$$
$$\underline{-6x-12}$$
$$17$$

The quotient is $x-3+\dfrac{17}{2x+4}$.

23. Using long division:

$$2a+3\overline{)6a^2+5a+1}$$

quotient: $3a-2$

$$\underline{6a^2+\ 9a}$$
$$-4a+1$$
$$\underline{-4a-6}$$
$$7$$

The quotient is $3a-2+\dfrac{7}{2a+3}$.

25. Using long division:

$$3a-5\overline{)6a^3-13a^2-4a+15}$$

quotient: $2a^2-a-3$

$$\underline{6a^3-10a^2}$$
$$-3a^2-4a$$
$$\underline{-3a^2+5a}$$
$$-9a+15$$
$$\underline{-9a+15}$$
$$0$$

The quotient is $2a^2-a-3$.

27. Using long division:

$$x+1\overline{)x^3+0x^2+4x+5}$$

quotient: x^2-x+5

$$\underline{x^3+\ x^2}$$
$$-x^2+4x$$
$$\underline{-x^2-\ x}$$
$$5x+5$$
$$\underline{5x+5}$$
$$0$$

The quotient is x^2-x+5.

29. Using long division:

$$x - 1 \overline{)x^3 + 0x^2 + 0x - 1}$$

quotient: $x^2 + x + 1$

$$\underline{x^3 - x^2}$$
$$x^2 + 0x$$
$$\underline{x^2 - x}$$
$$x - 1$$
$$\underline{x - 1}$$
$$0$$

The quotient is $x^2 + x + 1$.

31. Using long division:

$$x - 2 \overline{)x^3 + 0x^2 + 0x - 8}$$

quotient: $x^2 + 2x + 4$

$$\underline{x^3 - 2x^2}$$
$$2x^2 + 0x$$
$$\underline{2x^2 - 4x}$$
$$4x - 8$$
$$\underline{4x - 8}$$
$$0$$

The quotient is $x^2 + 2x + 4$.

33. The monthly payment is: $\dfrac{\$5{,}894}{12} \approx \$491.17$

35. The monthly payment is: $\dfrac{\$3{,}977}{12} \approx \$331.42$

37. Simplifying: $6(3 + 4) + 5 = 6(7) + 5 = 42 + 5 = 47$

39. Simplifying: $1^2 + 2^2 + 3^2 = 1 + 4 + 9 = 14$

41. Simplifying: $5(6 + 3 \bullet 2) + 4 + 3 \bullet 2 = 5(6 + 6) + 4 + 3 \bullet 2 = 5(12) + 4 + 3 \bullet 2 = 60 + 4 + 6 = 70$

43. Simplifying: $\left(1^3 + 2^3\right) + \left[(2 \bullet 3) + (4 \bullet 5)\right] = (1 + 8) + [6 + 20] = 9 + 26 = 35$

45. Simplifying: $[2 \bullet 3 + 4 + 5] \div 3 = (6 + 4 + 5) \div 3 = 15 \div 3 = 5$

47. Simplifying: $6 \bullet 10^3 + 5 \bullet 10^2 + 4 \bullet 0^1 = 6{,}000 + 500 + 40 = 6{,}540$

49. Simplifying: $1 \bullet 10^3 + 7 \bullet 10^2 + 6 \bullet 10^1 + 0 = 1{,}000 + 700 + 60 + 0 = 1{,}760$

51. Simplifying: $4 \bullet 2 - 1 + 5 \bullet 3 - 2 = 8 - 1 + 15 - 2 = 20$

53. Simplifying: $\left(2^3 + 3^2\right) \bullet 4 - 5 = (8 + 9) \bullet 4 - 5 = 17 \bullet 4 - 5 = 68 - 5 = 63$

55. Simplifying: $2\left(2^2 + 3^2\right) + 3\left(3^2\right) = 2(4 + 9) + 3(9) = 2(13) + 3(9) = 26 + 27 = 53$

Chapter 4 Test

See www.mathtv.com for video solutions to all problems in this chapter test.

Chapter 5
Factoring

5.1 The Greatest Common Factor and Factoring by Grouping

1. Factoring out the greatest common factor: $15x + 25 = 5(3x + 5)$

3. Factoring out the greatest common factor: $6a + 9 = 3(2a + 3)$

5. Factoring out the greatest common factor: $4x - 8y = 4(x - 2y)$

7. Factoring out the greatest common factor: $3x^2 - 6x - 9 = 3(x^2 - 2x - 3)$

9. Factoring out the greatest common factor: $3a^2 - 3a - 60 = 3(a^2 - a - 20)$

11. Factoring out the greatest common factor: $24y^2 - 52y + 24 = 4(6y^2 - 13y + 6)$

13. Factoring out the greatest common factor: $9x^2 - 8x^3 = x^2(9 - 8x)$

15. Factoring out the greatest common factor: $13a^2 - 26a^3 = 13a^2(1 - 2a)$

17. Factoring out the greatest common factor: $21x^2y - 28xy^2 = 7xy(3x - 4y)$

19. Factoring out the greatest common factor: $22a^2b^2 - 11ab^2 = 11ab^2(2a - 1)$

21. Factoring out the greatest common factor: $7x^3 + 21x^2 - 28x = 7x(x^2 + 3x - 4)$

23. Factoring out the greatest common factor: $121y^4 - 11x^4 = 11(11y^4 - x^4)$

25. Factoring out the greatest common factor: $100x^4 - 50x^3 + 25x^2 = 25x^2(4x^2 - 2x + 1)$

27. Factoring out the greatest common factor: $8a^2 + 16b^2 + 32c^2 = 8(a^2 + 2b^2 + 4c^2)$

29. Factoring out the greatest common factor: $4a^2b - 16ab^2 + 32a^2b^2 = 4ab(a - 4b + 8ab)$

31. Factoring out the greatest common factor: $121a^3b^2 - 22a^2b^3 + 33a^3b^3 = 11a^2b^2(11a - 2b + 3ab)$

33. Factoring out the greatest common factor: $12x^2y^3 - 72x^5y^3 - 36x^4y^4 = 12x^2y^3(1 - 6x^3 - 3x^2y)$

35. Factoring by grouping: $xy + 5x + 3y + 15 = x(y + 5) + 3(y + 5) = (y + 5)(x + 3)$

37. Factoring by grouping: $xy + 6x + 2y + 12 = x(y + 6) + 2(y + 6) = (y + 6)(x + 2)$

39. Factoring by grouping: $ab + 7a - 3b - 21 = a(b + 7) - 3(b + 7) = (b + 7)(a - 3)$

41. Factoring by grouping: $ax - bx + ay - by = x(a - b) + y(a - b) = (a - b)(x + y)$

43. Factoring by grouping: $2ax + 6x - 5a - 15 = 2x(a + 3) - 5(a + 3) = (a + 3)(2x - 5)$

45. Factoring by grouping: $3xb - 4b - 6x + 8 = b(3x - 4) - 2(3x - 4) = (3x - 4)(b - 2)$

47. Factoring by grouping: $x^2 + ax + 2x + 2a = x(x + a) + 2(x + a) = (x + a)(x + 2)$

49. Factoring by grouping: $x^2 - ax - bx + ab = x(x - a) - b(x - a) = (x - a)(x - b)$

51. Factoring by grouping:
$$ax + ay + bx + by + cx + cy = a(x+y) + b(x+y) + c(x+y) = (x+y)(a+b+c)$$

53. Factoring by grouping: $6x^2 + 9x + 4x + 6 = 3x(2x+3) + 2(2x+3) = (2x+3)(3x+2)$

55. Factoring by grouping: $20x^2 - 2x + 50x - 5 = 2x(10x-1) + 5(10x-1) = (10x-1)(2x+5)$

57. Factoring by grouping: $20x^2 + 4x + 25x + 5 = 4x(5x+1) + 5(5x+1) = (5x+1)(4x+5)$

59. Factoring by grouping: $x^3 + 2x^2 + 3x + 6 = x^2(x+2) + 3(x+2) = (x+2)(x^2+3)$

61. Factoring by grouping: $6x^3 - 4x^2 + 15x - 10 = 2x^2(3x-2) + 5(3x-2) = (3x-2)(2x^2+5)$

63. Its greatest common factor is $3 \cdot 2 = 6$.

65. The correct factoring is: $12x^2 + 6x + 3 = 3(4x^2 + 2x + 1)$

67. The factored form is: $A = 1000 + 1000r = 1000(1+r)$

Substituting $r = 0.12$: $A = 1000(1 + 0.12) = 1000(1.12) = \$1,120$

69. **a.**   Factoring: $A = 1,000,000 + 1,000,000r = 1,000,000(1+r)$

b.   Substituting $r = 0.30$: $A = 1,000,000(1 + 0.3) = 1,300,000$ bacteria

71. Multiplying using the FOIL method: $(x-7)(x+2) = x^2 - 7x + 2x - 14 = x^2 - 5x - 14$

73. Multiplying using the FOIL method: $(x-3)(x+2) = x^2 - 3x + 2x - 6 = x^2 - x - 6$

75. Multiplying using the column method: 77. Multiplying using the column method:

$$
\begin{array}{r}
x^2 - 3x + 9 \\
x + 3 \\
\hline
x^3 - 3x^2 + 9x \\
3x^2 - 9x + 27 \\
\hline
x^3 + 27
\end{array}
\qquad
\begin{array}{r}
x^2 + 4x - 3 \\
2x + 1 \\
\hline
2x^3 + 8x^2 - 6x \\
x^2 + 4x - 3 \\
\hline
2x^3 + 9x^2 - 2x - 3
\end{array}
$$

79. Multiplying: $3x^4(6x^3 - 4x^2 + 2x) = 3x^4 \cdot 6x^3 - 3x^4 \cdot 4x^2 + 3x^4 \cdot 2x = 18x^7 - 12x^6 + 6x^5$

81. Multiplying: $\left(x + \dfrac{1}{3}\right)\left(x + \dfrac{2}{3}\right) = x^2 + \dfrac{2}{3}x + \dfrac{1}{3}x + \dfrac{2}{9} = x^2 + x + \dfrac{2}{9}$

83. Multiplying: $(6x + 4y)(2x - 3y) = 12x^2 - 18xy + 8xy - 12y^2 = 12x^2 - 10xy - 12y^2$

85. Multiplying: $(9a + 1)(9a - 1) = 81a^2 - 9a + 9a - 1 = 81a^2 - 1$

87. Multiplying: $(x-9)(x-9) = x^2 - 9x - 9x + 81 = x^2 - 18x + 81$

89. Multiplying:
$$
\begin{aligned}
(x+2)(x^2 - 2x + 4) &= x(x^2 - 2x + 4) + 2(x^2 - 2x + 4) \\
&= x^3 - 2x^2 + 4x + 2x^2 - 4x + 8 \\
&= x^3 + 8
\end{aligned}
$$

5.2 Factoring Trinomials

1.   Factoring the trinomial: $x^2 + 7x + 12 = (x+3)(x+4)$

3.   Factoring the trinomial: $x^2 + 3x + 2 = (x+2)(x+1)$

5.   Factoring the trinomial: $a^2 + 10a + 21 = (a+7)(a+3)$

7.   Factoring the trinomial: $x^2 - 7x + 10 = (x-5)(x-2)$

9.   Factoring the trinomial: $y^2 - 10y + 21 = (y-7)(y-3)$

11. Factoring the trinomial: $x^2 - x - 12 = (x-4)(x+3)$

13. Factoring the trinomial: $y^2 + y - 12 = (y+4)(y-3)$

15. Factoring the trinomial: $x^2 + 5x - 14 = (x+7)(x-2)$

17. Factoring the trinomial: $r^2 - 8r - 9 = (r-9)(r+1)$

19. Factoring the trinomial: $x^2 - x - 30 = (x-6)(x+5)$

21. Factoring the trinomial: $a^2 + 15a + 56 = (a+7)(a+8)$

23. Factoring the trinomial: $y^2 - y - 42 = (y-7)(y+6)$

25. Factoring the trinomial: $x^2 + 13x + 42 = (x+7)(x+6)$

27. Factoring the trinomial: $2x^2 + 6x + 4 = 2(x^2 + 3x + 2) = 2(x+2)(x+1)$

29. Factoring the trinomial: $3a^2 - 3a - 60 = 3(a^2 - a - 20) = 3(a-5)(a+4)$

31. Factoring the trinomial: $100x^2 - 500x + 600 = 100(x^2 - 5x + 6) = 100(x-3)(x-2)$

33. Factoring the trinomial: $100p^2 - 1300p + 4000 = 100(p^2 - 13p + 40) = 100(p-8)(p-5)$

35. Factoring the trinomial: $x^4 - x^3 - 12x^2 = x^2(x^2 - x - 12) = x^2(x-4)(x+3)$

37. Factoring the trinomial: $2r^3 + 4r^2 - 30r = 2r(r^2 + 2r - 15) = 2r(r+5)(r-3)$

39. Factoring the trinomial: $2y^4 - 6y^3 - 8y^2 = 2y^2(y^2 - 3y - 4) = 2y^2(y-4)(y+1)$

41. Factoring the trinomial: $x^5 + 4x^4 + 4x^3 = x^3(x^2 + 4x + 4) = x^3(x+2)(x+2) = x^3(x+2)^2$

43. Factoring the trinomial: $3y^4 - 12y^3 - 15y^2 = 3y^2(y^2 - 4y - 5) = 3y^2(y-5)(y+1)$

45. Factoring the trinomial: $4x^4 - 52x^3 + 144x^2 = 4x^2(x^2 - 13x + 36) = 4x^2(x-9)(x-4)$

47. Factoring the trinomial: $x^2 + 5xy + 6y^2 = (x+2y)(x+3y)$

49. Factoring the trinomial: $x^2 - 9xy + 20y^2 = (x-4y)(x-5y)$

51. Factoring the trinomial: $a^2 + 2ab - 8b^2 = (a+4b)(a-2b)$

53. Factoring the trinomial: $a^2 - 10ab + 25b^2 = (a-5b)(a-5b) = (a-5b)^2$

55. Factoring the trinomial: $a^2 + 10ab + 25b^2 = (a+5b)(a+5b) = (a+5b)^2$

57. Factoring the trinomial: $x^2 + 2xa - 48a^2 = (x+8a)(x-6a)$

59. Factoring the trinomial: $x^2 - 5xb - 36b^2 = (x-9b)(x+4b)$

61. Factoring the trinomial: $x^4 - 5x^2 + 6 = (x^2 - 2)(x^2 - 3)$

63.   Factoring the trinomial: $x^2 - 80x - 2000 = (x - 100)(x + 20)$

65.   Factoring the trinomial: $x^2 - x + \dfrac{1}{4} = \left(x - \dfrac{1}{2}\right)\left(x - \dfrac{1}{2}\right) = \left(x - \dfrac{1}{2}\right)^2$

67.   Factoring the trinomial: $x^2 + 0.6x + 0.08 = (x + 0.4)(x + 0.2)$

69.   We can use long division to find the other factor:

$$\begin{array}{r} x + 16 \\ x + 8 \overline{\smash{\big)} x^2 + 24x + 128} \\ \underline{x^2 + 8x } \\ 16x + 128 \\ \underline{16x + 128} \\ 0 \end{array}$$

The other factor is $x + 16$.

71.   Using FOIL to multiply out the factors: $(4x + 3)(x - 1) = 4x^2 + 3x - 4x - 3 = 4x^2 - x - 3$

73.   Multiplying using the FOIL method: $(6a + 1)(a + 2) = 6a^2 + a + 12a + 2 = 6a^2 + 13a + 2$

75.   Multiplying using the FOIL method: $(3a + 2)(2a + 1) = 6a^2 + 4a + 3a + 2 = 6a^2 + 7a + 2$

77.   Multiplying using the FOIL method: $(6a + 2)(a + 1) = 6a^2 + 2a + 6a + 2 = 6a^2 + 8a + 2$

5.3 More Trinomials to Factor

1.   Factoring the trinomial: $2x^2 + 7x + 3 = (2x + 1)(x + 3)$

3.   Factoring the trinomial: $2a^2 - a - 3 = (2a - 3)(a + 1)$

5.   Factoring the trinomial: $3x^2 + 2x - 5 = (3x + 5)(x - 1)$

7.   Factoring the trinomial: $3y^2 - 14y - 5 = (3y + 1)(y - 5)$

9.   Factoring the trinomial: $6x^2 + 13x + 6 = (3x + 2)(2x + 3)$

11.   Factoring the trinomial: $4x^2 - 12xy + 9y^2 = (2x - 3y)(2x - 3y) = (2x - 3y)^2$

13.   Factoring the trinomial: $4y^2 - 11y - 3 = (4y + 1)(y - 3)$

15.   Factoring the trinomial: $20x^2 - 41x + 20 = (4x - 5)(5x - 4)$

17.   Factoring the trinomial: $20a^2 + 48ab - 5b^2 = (10a - b)(2a + 5b)$

19.   Factoring the trinomial: $20x^2 - 21x - 5 = (4x - 5)(5x + 1)$

21.   Factoring the trinomial: $12m^2 + 16m - 3 = (6m - 1)(2m + 3)$

23.   Factoring the trinomial: $20x^2 + 37x + 15 = (4x + 5)(5x + 3)$

25.   Factoring the trinomial: $12a^2 - 25ab + 12b^2 = (3a - 4b)(4a - 3b)$

27.   Factoring the trinomial: $3x^2 - xy - 14y^2 = (3x - 7y)(x + 2y)$

29.   Factoring the trinomial: $14x^2 + 29x - 15 = (2x + 5)(7x - 3)$

31.   Factoring the trinomial: $6x^2 - 43x + 55 = (3x - 5)(2x - 11)$

33. Factoring the trinomial: $15t^2 - 67t + 38 = (5t - 19)(3t - 2)$

35. Factoring the trinomial: $4x^2 + 2x - 6 = 2(2x^2 + x - 3) = 2(2x + 3)(x - 1)$

37. Factoring the trinomial: $24a^2 - 50a + 24 = 2(12a^2 - 25a + 12) = 2(4a - 3)(3a - 4)$

39. Factoring the trinomial: $10x^3 - 23x^2 + 12x = x(10x^2 - 23x + 12) = x(5x - 4)(2x - 3)$

41. Factoring the trinomial: $6x^4 - 11x^3 - 10x^2 = x^2(6x^2 - 11x - 10) = x^2(3x + 2)(2x - 5)$

43. Factoring the trinomial: $10a^3 - 6a^2 - 4a = 2a(5a^2 - 3a - 2) = 2a(5a + 2)(a - 1)$

45. Factoring the trinomial: $15x^3 - 102x^2 - 21x = 3x(5x^2 - 34x - 7) = 3x(5x + 1)(x - 7)$

47. Factoring the trinomial: $35y^3 - 60y^2 - 20y = 5y(7y^2 - 12y - 4) = 5y(7y + 2)(y - 2)$

49. Factoring the trinomial: $15a^4 - 2a^3 - a^2 = a^2(15a^2 - 2a - 1) = a^2(5a + 1)(3a - 1)$

51. Factoring the trinomial: $24x^2y - 6xy - 45y = 3y(8x^2 - 2x - 15) = 3y(4x + 5)(2x - 3)$

53. Factoring the trinomial: $12x^2y - 34xy^2 + 14y^3 = 2y(6x^2 - 17xy + 7y^2) = 2y(2x - y)(3x - 7y)$

55. Evaluating each expression when $x = 2$:
$$2x^2 + 7x + 3 = 2(2)^2 + 7(2) + 3 = 8 + 14 + 3 = 25$$
$$(2x + 1)(x + 3) = (2 \bullet 2 + 1)(2 + 3) = (5)(5) = 25$$

57. Multiplying using the difference of squares formula: $(2x + 3)(2x - 3) = (2x)^2 - (3)^2 = 4x^2 - 9$

59. Multiplying using the difference of squares formula:
$$(x + 3)(x - 3)(x^2 + 9) = (x^2 - 9)(x^2 + 9) = (x^2)^2 - (9)^2 = x^4 - 81$$

61. Factoring: $h = 8 + 62t - 16t^2 = 2(4 + 31t - 8t^2) = 2(4 - t)(1 + 8t)$. Now completing the table:

Time t (seconds)	0	1	2	3	4
Height h (feet)	8	54	68	50	0

63. a. Factoring: $V = 99x - 40x^2 + 4x^3 = x(99 - 40x + 4x^2) = x(9 - 2x)(11 - 2x)$
   b. The original dimensions were 9 inches by 11 inches.

65. Multiplying: $(x + 3)(x - 3) = x^2 - (3)^2 = x^2 - 9$

67. Multiplying: $(x + 5)(x - 5) = x^2 - (5)^2 = x^2 - 25$

69. Multiplying: $(x + 7)(x - 7) = x^2 - (7)^2 = x^2 - 49$

71. Multiplying: $(x + 9)(x - 9) = x^2 - (9)^2 = x^2 - 81$

73. Multiplying: $(2x - 3y)(2x + 3y) = (2x)^2 - (3y)^2 = 4x^2 - 9y^2$

75. Multiplying: $(x^2 + 4)(x + 2)(x - 2) = (x^2 + 4)(x^2 - 4) = (x^2)^2 - (4)^2 = x^4 - 16$

77. Multiplying: $(x + 3)^2 = x^2 + 2(x)(3) + (3)^2 = x^2 + 6x + 9$

79. Multiplying: $(x + 5)^2 = x^2 + 2(x)(5) + (5)^2 = x^2 + 10x + 25$

81. Multiplying: $(x+7)^2 = x^2 + 2(x)(7) + (7)^2 = x^2 + 14x + 49$

83. Multiplying: $(x+9)^2 = x^2 + 2(x)(9) + (9)^2 = x^2 + 18x + 81$

85. Multiplying: $(2x+3)^2 = (2x)^2 + 2(2x)(3) + (3)^2 = 4x^2 + 12x + 9$

87. Multiplying: $(4x-2y)^2 = (4x)^2 - 2(4x)(2y) + (2y)^2 = 16x^2 - 16xy + 4y^2$

5.4 The Difference of Two Squares

1.   Factoring the binomial: $x^2 - 9 = (x+3)(x-3)$

3.   Factoring the binomial: $a^2 - 36 = (a+6)(a-6)$

5.   Factoring the binomial: $x^2 - 49 = (x+7)(x-7)$

7.   Factoring the binomial: $4a^2 - 16 = 4(a^2 - 4) = 4(a+2)(a-2)$

9.   The expression $9x^2 + 25$ cannot be factored.

11. Factoring the binomial: $25x^2 - 169 = (5x+13)(5x-13)$

13. Factoring the binomial: $9a^2 - 16b^2 = (3a+4b)(3a-4b)$

15. Factoring the binomial: $9 - m^2 = (3+m)(3-m)$

17. Factoring the binomial: $25 - 4x^2 = (5+2x)(5-2x)$

19. Factoring the binomial: $2x^2 - 18 = 2(x^2 - 9) = 2(x+3)(x-3)$

21. Factoring the binomial: $32a^2 - 128 = 32(a^2 - 4) = 32(a+2)(a-2)$

23. Factoring the binomial: $8x^2y - 18y = 2y(4x^2 - 9) = 2y(2x+3)(2x-3)$

25. Factoring the binomial: $a^4 - b^4 = (a^2 + b^2)(a^2 - b^2) = (a^2 + b^2)(a+b)(a-b)$

27. Factoring the binomial: $16m^4 - 81 = (4m^2 + 9)(4m^2 - 9) = (4m^2 + 9)(2m+3)(2m-3)$

29. Factoring the binomial: $3x^3y - 75xy^3 = 3xy(x^2 - 25y^2) = 3xy(x+5y)(x-5y)$

31. Factoring the trinomial: $x^2 - 2x + 1 = (x-1)(x-1) = (x-1)^2$

33. Factoring the trinomial: $x^2 + 2x + 1 = (x+1)(x+1) = (x+1)^2$

35. Factoring the trinomial: $a^2 - 10a + 25 = (a-5)(a-5) = (a-5)^2$

37. Factoring the trinomial: $y^2 + 4y + 4 = (y+2)(y+2) = (y+2)^2$

39. Factoring the trinomial: $x^2 - 4x + 4 = (x-2)(x-2) = (x-2)^2$

41. Factoring the trinomial: $m^2 - 12m + 36 = (m-6)(m-6) = (m-6)^2$

43. Factoring the trinomial: $4a^2 + 12a + 9 = (2a+3)(2a+3) = (2a+3)^2$

45. Factoring the trinomial: $49x^2 - 14x + 1 = (7x-1)(7x-1) = (7x-1)^2$

47. Factoring the trinomial: $9y^2 - 30y + 25 = (3y-5)(3y-5) = (3y-5)^2$

49. Factoring the trinomial: $x^2 + 10xy + 25y^2 = (x+5y)(x+5y) = (x+5y)^2$

51. Factoring the trinomial: $9a^2 + 6ab + b^2 = (3a+b)(3a+b) = (3a+b)^2$

53. Factoring the trinomial: $3a^2 + 18a + 27 = 3(a^2 + 6a + 9) = 3(a+3)(a+3) = 3(a+3)^2$

55. Factoring the trinomial:
$$2x^2 + 20xy + 50y^2 = 2(x^2 + 10xy + 25y^2) = 2(x+5y)(x+5y) = 2(x+5y)^2$$

57. Factoring the trinomial:
$$5x^3 + 30x^2y + 45xy^2 = 5x(x^2 + 6xy + 9y^2) = 5x(x+3y)(x+3y) = 5x(x+3y)^2$$

59. Factoring by grouping: $x^2 + 6x + 9 - y^2 = (x+3)^2 - y^2 = (x+3+y)(x+3-y)$

61. Factoring by grouping: $x^2 + 2xy + y^2 - 9 = (x+y)^2 - 9 = (x+y+3)(x+y-3)$

63. Since $(x+7)^2 = x^2 + 14x + 49$, the value is $b = 14$.

65. Since $(x+5)^2 = x^2 + 10x + 25$, the value is $c = 25$.

67. **a.** Subtracting square areas, the area is $x^2 - 4^2 = x^2 - 16$.

    b. Factoring: $x^2 - 16 = (x+4)(x-4)$

    c. The square can be rearranged to be a rectangle with dimensions $x - 4$ by $x + 4$. Cut off the right "flap", then place it along the bottom of the left rectangle to produce the desired rectangle.

69. The area is $a^2 - b^2 = (a+b)(a-b)$.

71. **a.** Multiplying: $1^3 = 1$ **b.** Multiplying: $2^3 = 8$

    c. Multiplying: $3^3 = 27$ **d.** Multiplying: $4^3 = 64$

    e. Multiplying: $5^3 = 125$

73. **a.** Multiplying: $x(x^2 - x + 1) = x^3 - x^2 + x$

    b. Multiplying: $1(x^2 - x + 1) = x^2 - x + 1$

    c. Multiplying:
$$(x+1)(x^2 - x + 1) = x(x^2 - x + 1) + 1(x^2 - x + 1) = x^3 - x^2 + x + x^2 - x + 1 = x^3 + 1$$

75. **a.** Multiplying: $x(x^2 - 2x + 4) = x^3 - 2x^2 + 4x$

    b. Multiplying: $2(x^2 - 2x + 4) = 2x^2 - 4x + 8$

    c. Multiplying:
$$\begin{aligned}(x+2)(x^2 - 2x + 4) &= x(x^2 - 2x + 4) + 2(x^2 - 2x + 4) \\ &= x^3 - 2x^2 + 4x + 2x^2 - 4x + 8 \\ &= x^3 + 8\end{aligned}$$

77. **a.** Multiplying: $x(x^2 - 3x + 9) = x^3 - 3x^2 + 9x$

    b. Multiplying: $3(x^2 - 3x + 9) = 3x^2 - 9x + 27$

    c. Multiplying:
$$\begin{aligned}(x+3)(x^2 - 3x + 9) &= x(x^2 - 3x + 9) + 3(x^2 - 3x + 9) \\ &= x^3 - 3x^2 + 9x + 3x^2 - 9x + 27 \\ &= x^3 + 27\end{aligned}$$

5.5 The Sum and Difference of Two Cubes

1. Factoring: $x^3 - y^3 = (x - y)(x^2 + xy + y^2)$

3. Factoring: $a^3 + 8 = (a + 2)(a^2 - 2a + 4)$

5. Factoring: $27 + x^3 = (3 + x)(9 - 3x + x^2)$

7. Factoring: $y^3 - 1 = (y - 1)(y^2 + y + 1)$

9. Factoring: $y^3 - 64 = (y - 4)(y^2 + 4y + 16)$

11. Factoring: $125h^3 - t^3 = (5h - t)(25h^2 + 5ht + t^2)$

13. Factoring: $x^3 - 216 = (x - 6)(x^2 + 6x + 36)$

15. Factoring: $2y^3 - 54 = 2(y^3 - 27) = 2(y - 3)(y^2 + 3y + 9)$

17. Factoring: $2a^3 - 128b^3 = 2(a^3 - 64b^3) = 2(a - 4b)(a^2 + 4ab + 16b^2)$

19. Factoring: $2x^3 + 432y^3 = 2(x^3 + 216y^3) = 2(x + 6y)(x^2 - 6xy + 36y^2)$

21. Factoring: $10a^3 - 640b^3 = 10(a^3 - 64b^3) = 10(a - 4b)(a^2 + 4ab + 16b^2)$

23. Factoring: $10r^3 - 1250 = 10(r^3 - 125) = 10(r - 5)(r^2 + 5r + 25)$

25. Factoring: $64 + 27a^3 = (4 + 3a)(16 - 12a + 9a^2)$

27. Factoring: $8x^3 - 27y^3 = (2x - 3y)(4x^2 + 6xy + 9y^2)$

29. Factoring: $t^3 + \dfrac{1}{27} = \left(t + \dfrac{1}{3}\right)\left(t^2 - \dfrac{1}{3}t + \dfrac{1}{9}\right)$

31. Factoring: $27x^3 - \dfrac{1}{27} = \left(3x - \dfrac{1}{3}\right)\left(9x^2 + x + \dfrac{1}{9}\right)$

33. Factoring: $64a^3 + 125b^3 = (4a + 5b)(16a^2 - 20ab + 25b^2)$

35. Factoring: $\dfrac{1}{8}x^3 - \dfrac{1}{27}y^3 = \left(\dfrac{1}{2}x - \dfrac{1}{3}y\right)\left(\dfrac{1}{4}x^2 + \dfrac{1}{6}xy + \dfrac{1}{9}y^2\right)$

37. Factoring: $a^6 - b^6 = (a^3 + b^3)(a^3 - b^3) = (a + b)(a^2 - ab + b^2)(a - b)(a^2 + ab + b^2)$

39. Factoring:
$$64x^6 - y^6 = (8x^3 + y^3)(8x^3 - y^3) = (2x + y)(4x^2 - 2xy + y^2)(2x - y)(4x^2 + 2xy + y^2)$$

41. Factoring:
$$x^6 - (5y)^6 = \left(x^3 + (5y)^3\right)\left(x^3 - (5y)^3\right)$$
$$= (x + 5y)(x^2 - 5xy + 25y^2)(x - 5y)(x^2 + 5xy + 25y^2)$$

43. Multiplying: $2x^3(x + 2)(x - 2) = 2x^3(x^2 - 4) = 2x^5 - 8x^3$

45. Multiplying: $3x^2(x - 3)^2 = 3x^2(x^2 - 6x + 9) = 3x^4 - 18x^3 + 27x^2$

47. Multiplying: $y(y^2 + 25) = y^3 + 25y$

49. Multiplying: $(5a - 2)(3a + 1) = 15a^2 + 5a - 6a - 2 = 15a^2 - a - 2$

51. Multiplying: $4x^2(x - 5)(x + 2) = 4x^2(x^2 - 3x - 10) = 4x^4 - 12x^3 - 40x^2$

53. Multiplying: $2ab^3(b^2 - 4b + 1) = 2ab^5 - 8ab^4 + 2ab^3$

5.6 Factoring: A General Review

1. Factoring the polynomial: $x^2 - 81 = (x+9)(x-9)$

3. Factoring the polynomial: $x^2 + 2x - 15 = (x+5)(x-3)$

5. Factoring the polynomial: $x^2 + 6x + 9 = (x+3)(x+3) = (x+3)^2$

7. Factoring the polynomial: $y^2 - 10y + 25 = (y-5)(y-5) = (y-5)^2$

9. Factoring the polynomial: $2a^3b + 6a^2b + 2ab = 2ab(a^2 + 3a + 1)$

11. The polynomial $x^2 + x + 1$ cannot be factored.

13. Factoring the polynomial: $12a^2 - 75 = 3(4a^2 - 25) = 3(2a+5)(2a-5)$

15. Factoring the polynomial: $9x^2 - 12xy + 4y^2 = (3x-2y)(3x-2y) = (3x-2y)^2$

17. Factoring the polynomial: $4x^3 + 16xy^2 = 4x(x^2 + 4y^2)$

19. Factoring the polynomial: $2y^3 + 20y^2 + 50y = 2y(y^2 + 10y + 25) = 2y(y+5)(y+5) = 2y(y+5)^2$

21. Factoring the polynomial: $a^6 + 4a^4b^2 = a^4(a^2 + 4b^2)$

23. Factoring the polynomial: $xy + 3x + 4y + 12 = x(y+3) + 4(y+3) = (y+3)(x+4)$

25. Factoring the polynomial: $x^4 - 16 = (x^2 + 4)(x^2 - 4) = (x^2 + 4)(x+2)(x-2)$

27. Factoring the polynomial: $xy - 5x + 2y - 10 = x(y-5) + 2(y-5) = (y-5)(x+2)$

29. Factoring the polynomial: $5a^2 + 10ab + 5b^2 = 5(a^2 + 2ab + b^2) = 5(a+b)(a+b) = 5(a+b)^2$

31. The polynomial $x^2 + 49$ cannot be factored.

33. Factoring the polynomial: $3x^2 + 15xy + 18y^2 = 3(x^2 + 5xy + 6y^2) = 3(x+2y)(x+3y)$

35. Factoring the polynomial: $2x^2 + 15x - 38 = (2x+19)(x-2)$

37. Factoring the polynomial: $100x^2 - 300x + 200 = 100(x^2 - 3x + 2) = 100(x-2)(x-1)$

39. Factoring the polynomial: $x^2 - 64 = (x+8)(x-8)$

41. Factoring the polynomial: $x^2 + 3x + ax + 3a = x(x+3) + a(x+3) = (x+3)(x+a)$

43. Factoring the polynomial: $49a^7 - 9a^5 = a^5(49a^2 - 9) = a^5(7a+3)(7a-3)$

45. The polynomial $49x^2 + 9y^2$ cannot be factored.

47. Factoring the polynomial: $25a^3 + 20a^2 + 3a = a(25a^2 + 20a + 3) = a(5a+3)(5a+1)$

49. Factoring the polynomial: $xa - xb + ay - by = x(a-b) + y(a-b) = (a-b)(x+y)$

51. Factoring the polynomial: $48a^4b - 3a^2b = 3a^2b(16a^2 - 1) = 3a^2b(4a+1)(4a-1)$

53. Factoring the polynomial: $20x^4 - 45x^2 = 5x^2(4x^2 - 9) = 5x^2(2x+3)(2x-3)$

55. Factoring the polynomial: $3x^2 + 35xy - 82y^2 = (3x+41y)(x-2y)$

57. Factoring the polynomial: $16x^5 - 44x^4 + 30x^3 = 2x^3(8x^2 - 22x + 15) = 2x^3(2x-3)(4x-5)$

59. Factoring the polynomial: $2x^2 + 2ax + 3x + 3a = 2x(x+a) + 3(x+a) = (x+a)(2x+3)$

61.   Factoring the polynomial: $y^4 - 1 = (y^2 + 1)(y^2 - 1) = (y^2 + 1)(y + 1)(y - 1)$

63.   Factoring the polynomial:

$$12x^4y^2 + 36x^3y^3 + 27x^2y^4 = 3x^2y^2 (4x^2 + 12xy + 9y^2)$$

$$= 3x^2y^2 (2x + 3y)(2x + 3y)$$

$$= 3x^2y^2 (2x + 3y)^2$$

65.   Solving the equation:

$$3x - 6 = 9$$

$$3x = 15$$

$$x = 5$$

67.   Solving the equation:

$$2x + 3 = 0$$

$$2x = -3$$

$$x = -\frac{3}{2}$$

69.   Solving the equation:

$$4x + 3 = 0$$

$$4x = -3$$

$$x = -\frac{3}{4}$$

5.7 Solving Equations by Factoring

1.   Setting each factor equal to 0:

$$x + 2 = 0 \qquad\qquad x - 1 = 0$$

$$x = -2 \qquad\qquad\quad x = 1$$

The solutions are –2 and 1.

3.   Setting each factor equal to 0:

$$a - 4 = 0 \qquad\qquad a - 5 = 0$$

$$a = 4 \qquad\qquad\quad a = 5$$

The solutions are 4 and 5.

5.   Setting each factor equal to 0:

$$x = 0 \qquad\qquad x + 1 = 0 \qquad\qquad x - 3 = 0$$

$$x = -1 \qquad\qquad\quad x = 3$$

The solutions are 0, –1 and 3.

7.   Setting each factor equal to 0:

$$3x + 2 = 0 \qquad\qquad 2x + 3 = 0$$

$$3x = -2 \qquad\qquad\quad 2x = -3$$

$$x = -\frac{2}{3} \qquad\qquad\quad x = -\frac{3}{2}$$

The solutions are $-\dfrac{2}{3}$ and $-\dfrac{3}{2}$.

9. Setting each factor equal to 0:

$$m = 0 \qquad 3m + 4 = 0 \qquad\qquad 3m - 4 = 0$$
$$3m = -4 \qquad\qquad 3m = 4$$
$$m = -\frac{4}{3} \qquad\qquad m = \frac{4}{3}$$

The solutions are 0, $-\frac{4}{3}$ and $\frac{4}{3}$.

11. Setting each factor equal to 0:

$$2y = 0 \qquad 3y + 1 = 0 \qquad\qquad 5y + 3 = 0$$
$$y = 0 \qquad 3y = -1 \qquad\qquad 5y = -3$$
$$y = -\frac{1}{3} \qquad\qquad y = -\frac{3}{5}$$

The solutions are 0, $-\frac{1}{3}$ and $-\frac{3}{5}$.

13. Solving by factoring:

$$x^2 + 3x + 2 = 0$$
$$(x + 2)(x + 1) = 0$$
$$x = -2, -1$$

15. Solving by factoring:

$$x^2 - 9x + 20 = 0$$
$$(x - 4)(x - 5) = 0$$
$$x = 4, 5$$

17. Solving by factoring:

$$a^2 - 2a - 24 = 0$$
$$(a - 6)(a + 4) = 0$$
$$a = -4, 6$$

19. Solving by factoring:

$$100x^2 - 500x + 600 = 0$$
$$100\left(x^2 - 5x + 6\right) = 0$$
$$100(x - 2)(x - 3) = 0$$
$$x = 2, 3$$

21. Solving by factoring:

$$x^2 = -6x - 9$$
$$x^2 + 6x + 9 = 0$$
$$(x + 3)^2 = 0$$
$$x + 3 = 0$$
$$x = -3$$

23. Solving by factoring:

$$a^2 - 16 = 0$$
$$(a + 4)(a - 4) = 0$$
$$a = -4, 4$$

25. Solving by factoring:

$$2x^2 + 5x - 12 = 0$$
$$(2x - 3)(x + 4) = 0$$
$$x = \frac{3}{2}, -4$$

27. Solving by factoring:

$$9x^2 + 12x + 4 = 0$$
$$(3x + 2)^2 = 0$$
$$3x + 2 = 0$$
$$x = -\frac{2}{3}$$

29.   Solving by factoring:
$$a^2 + 25 = 10a$$
$$a^2 - 10a + 25 = 0$$
$$(a-5)^2 = 0$$
$$a - 5 = 0$$
$$a = 5$$

31.   Solving by factoring:
$$2x^2 = 3x + 20$$
$$2x^2 - 3x - 20 = 0$$
$$(2x+5)(x-4) = 0$$
$$x = -\frac{5}{2}, 4$$

33.   Solving by factoring:
$$3m^2 = 20 - 7m$$
$$3m^2 + 7m - 20 = 0$$
$$(3m-5)(m+4) = 0$$
$$m = \frac{5}{3}, -4$$

35.   Solving by factoring:
$$4x^2 - 49 = 0$$
$$(2x+7)(2x-7) = 0$$
$$x = -\frac{7}{2}, \frac{7}{2}$$

37.   Solving by factoring:
$$x^2 + 6x = 0$$
$$x(x+6) = 0$$
$$x = 0, -6$$

39.   Solving by factoring:
$$x^2 - 3x = 0$$
$$x(x-3) = 0$$
$$x = 0, 3$$

41.   Solving by factoring:
$$2x^2 = 8x$$
$$2x^2 - 8x = 0$$
$$2x(x-4) = 0$$
$$x = 0, 4$$

43.   Solving by factoring:
$$3x^2 = 15x$$
$$3x^2 - 15x = 0$$
$$3x(x-5) = 0$$
$$x = 0, 5$$

45.   Solving by factoring:
$$1,400 = 400 + 700x - 100x^2$$
$$100x^2 - 700x + 1,000 = 0$$
$$100(x^2 - 7x + 10) = 0$$
$$100(x-5)(x-2) = 0$$
$$x = 2, 5$$

47.   Solving by factoring:
$$6x^2 = -5x + 4$$
$$6x^2 + 5x - 4 = 0$$
$$(3x+4)(2x-1) = 0$$
$$x = -\frac{4}{3}, \frac{1}{2}$$

49.   Solving by factoring:
$$x(2x-3) = 20$$
$$2x^2 - 3x = 20$$
$$2x^2 - 3x - 20 = 0$$
$$(2x+5)(x-4) = 0$$
$$x = -\frac{5}{2}, 4$$

51. Solving by factoring:
$$t(t+2) = 80$$
$$t^2 + 2t = 80$$
$$t^2 + 2t - 80 = 0$$
$$(t+10)(t-8) = 0$$
$$t = -10, 8$$

53. Solving by factoring:
$$4000 = (1300 - 100p)p$$
$$4000 = 1300p - 100p^2$$
$$100p^2 - 1300p + 4000 = 0$$
$$100(p^2 - 13p + 40) = 0$$
$$100(p-8)(p-5) = 0$$
$$p = 5, 8$$

55. Solving by factoring:
$$x(14 - x) = 48$$
$$14x - x^2 = 48$$
$$-x^2 + 14x - 48 = 0$$
$$x^2 - 14x + 48 = 0$$
$$(x-6)(x-8) = 0$$
$$x = 6, 8$$

57. Solving by factoring:
$$(x+5)^2 = 2x + 9$$
$$x^2 + 10x + 25 = 2x + 9$$
$$x^2 + 8x + 16 = 0$$
$$(x+4)^2 = 0$$
$$x + 4 = 0$$
$$x = -4$$

59. Solving by factoring:

$$(y-6)^2 = y - 4$$
$$y^2 - 12y + 36 = y - 4$$
$$y^2 - 13y + 40 = 0$$
$$(y-5)(y-8) = 0$$
$$y = 5, 8$$

61. Solving by factoring:
$$10^2 = (x+2)^2 + x^2$$
$$100 = x^2 + 4x + 4 + x^2$$
$$100 = 2x^2 + 4x + 4$$
$$0 = 2x^2 + 4x - 96$$
$$0 = 2(x^2 + 2x - 48)$$
$$0 = 2(x+8)(x-6)$$
$$x = -8, 6$$

63. Solving by factoring:
$$2x^3 + 11x^2 + 12x = 0$$
$$x(2x^2 + 11x + 12) = 0$$
$$x(2x+3)(x+4) = 0$$
$$x = 0, -\frac{3}{2}, -4$$

65. Solving by factoring:
$$4y^3 - 2y^2 - 30y = 0$$
$$2y(2y^2 - y - 15) = 0$$
$$2y(2y+5)(y-3) = 0$$
$$y = 0, -\frac{5}{2}, 3$$

67. Solving by factoring:
$$8x^3 + 16x^2 = 10x$$
$$8x^3 + 16x^2 - 10x = 0$$
$$2x(4x^2 + 8x - 5) = 0$$
$$2x(2x-1)(2x+5) = 0$$
$$x = 0, \frac{1}{2}, -\frac{5}{2}$$

69. Solving by factoring:
$$20a^3 = -18a^2 + 18a$$
$$20a^3 + 18a^2 - 18a = 0$$
$$2a(10a^2 + 9a - 9) = 0$$
$$2a(5a-3)(2a+3) = 0$$
$$a = 0, \frac{3}{5}, -\frac{3}{2}$$

71. Solving by factoring:
$$16t^2 - 32t + 12 = 0$$
$$4(4t^2 - 8t + 3) = 0$$
$$4(2t - 1)(2t - 3) = 0$$
$$t = \frac{1}{2}, \frac{3}{2}$$

73. Solving by factoring:
$$(a-5)(a+4) = -2a$$
$$a^2 - a - 20 = -2a$$
$$a^2 + a - 20 = 0$$
$$(a+5)(a-4) = 0$$
$$a = -5, 4$$

75. Solving the equation:
$$3x(x+1) - 2x(x-5) = -42$$
$$3x^2 + 3x - 2x^2 + 10x = -42$$
$$x^2 + 13x + 42 = 0$$
$$(x+7)(x+6) = 0$$
$$x = -7, -6$$

77. Solving the equation:
$$2x(x+3) = x(x+2) - 3$$
$$2x^2 + 6x = x^2 + 2x - 3$$
$$x^2 + 4x + 3 = 0$$
$$(x+3)(x+1) = 0$$
$$x = -3, -1$$

79. Solving the equation:

$$a(a-3) + 6 = 2a$$
$$a^2 - 3a + 6 = 2a$$
$$a^2 - 5a + 6 = 0$$
$$(a-2)(a-3) = 0$$
$$a = 2, 3$$

81. Solving the equation:
$$15(x+20) + 15x = 2x(x+20)$$
$$15x + 300 + 15x = 2x^2 + 40x$$
$$30x + 300 = 2x^2 + 40x$$
$$0 = 2x^2 + 10x - 300$$
$$0 = 2(x^2 + 5x - 150)$$
$$0 = 2(x+15)(x-10)$$
$$x = -15, 10$$

83. Solving the equation:
$$15 = a(a+2)$$
$$15 = a^2 + 2a$$
$$0 = a^2 + 2a - 15$$
$$0 = (a+5)(a-3)$$
$$a = -5, 3$$

85. Solving the equation:
$$x^3 + 3x^2 - 4x - 12 = 0$$
$$x^2(x+3) - 4(x+3) = 0$$
$$(x+3)(x^2 - 4) = 0$$
$$(x+3)(x+2)(x-2) = 0$$
$$x = -3, -2, 2$$

87. Solving by factoring:
$$x^3 + x^2 - 16x - 16 = 0$$
$$x^2(x+1) - 16(x+1) = 0$$
$$(x+1)(x^2 - 16) = 0$$
$$(x+1)(x+4)(x-4) = 0$$
$$x = -1, -4, 4$$

89. Let x and $x + 1$ represent the two consecutive integers. The equation is: $x(x+1) = 72$

91. Let x and $x + 2$ represent the two consecutive odd integers. The equation is: $x(x+2) = 99$

93. Let x and $x + 2$ represent the two consecutive even integers. The equation is:
$$x(x+2) = 5\left[x + (x+2)\right] - 10$$

95.   Let x represent the cost of the suit and $5x$ represent the cost of the bicycle. The equation is:
$$x + 5x = 90$$
$$6x = 90$$
$$x = 15$$
$$5x = 75$$
The suit costs $15 and the bicycle costs $75.

97.   Let x represent the cost of the lot and $4x$ represent the cost of the house. The equation is:
$$x + 4x = 3000$$
$$5x = 3000$$
$$x = 600$$
$$4x = 2400$$
The lot costs $600 and the house costs $2,400.

5.8 Applications

1.   Let x and $x + 2$ represent the two integers. The equation is:
$$x(x + 2) = 80$$
$$x^2 + 2x = 80$$
$$x^2 + 2x - 80 = 0$$
$$(x + 10)(x - 8) = 0$$
$$x = -10, 8$$
$$x + 2 = -8, 10$$
The two numbers are either −10 and −8, or 8 and 10.

3.   Let x and $x + 2$ represent the two integers. The equation is:
$$x(x + 2) = 99$$
$$x^2 + 2x = 99$$
$$x^2 + 2x - 99 = 0$$
$$(x + 11)(x - 9) = 0$$
$$x = -11, 9$$
$$x + 2 = -9, 11$$
The two numbers are either −11 and −9, or 9 and 11.

5.   Let x and $x + 2$ represent the two integers. The equation is:
$$x(x + 2) = 5(x + x + 2) - 10$$
$$x^2 + 2x = 5(2x + 2) - 10$$
$$x^2 + 2x = 10x + 10 - 10$$
$$x^2 + 2x = 10x$$
$$x^2 - 8x = 0$$
$$x(x - 8) = 0$$
$$x = 0, 8$$
$$x + 2 = 2, 10$$
The two numbers are either 0 and 2, or 8 and 10.

7.   Let x and $14 - x$ represent the two numbers. The equation is:
$$x(14 - x) = 48$$
$$14x - x^2 = 48$$
$$0 = x^2 - 14x + 48$$
$$0 = (x - 8)(x - 6)$$
$$x = 8, 6$$
$$14 - x = 6, 8$$
The two numbers are 6 and 8.

9.   Let x and $5x + 2$ represent the two numbers. The equation is:
$$x(5x + 2) = 24$$
$$5x^2 + 2x = 24$$
$$5x^2 + 2x - 24 = 0$$
$$(5x + 12)(x - 2) = 0$$
$$x = -\frac{12}{5}, 2$$
$$5x + 2 = -10, 12$$
The two numbers are either $-\frac{12}{5}$ and -10, or 2 and 12.

11.   Let x and $4x$ represent the two numbers. The equation is:
$$x(4x) = 4(x + 4x)$$
$$4x^2 = 4(5x)$$
$$4x^2 = 20x$$
$$4x^2 - 20x = 0$$
$$4x(x - 5) = 0$$
$$x = 0, 5$$
$$4x = 0, 20$$
The two numbers are either 0 and 0, or 5 and 20.

13.   Let w represent the width and $w + 1$ represent the length. The equation is:
$$w(w + 1) = 12$$
$$w^2 + w = 12$$
$$w^2 + w - 12 = 0$$
$$(w + 4)(w - 3) = 0$$
$$w = 3 \quad (w = -4 \text{ is impossible})$$
$$w + 1 = 4$$
The width is 3 inches and the length is 4 inches.

15. Let b represent the base and $2b$ represent the height. The equation is:

$$\frac{1}{2}b(2b) = 9$$

$$b^2 = 9$$

$$b^2 - 9 = 0$$

$$(b+3)(b-3) = 0$$

$$b = 3 \quad (b = -3 \text{ is impossible})$$

The base is 3 inches.

17. Let x and $x + 2$ represent the two legs. The equation is:

$$x^2 + (x+2)^2 = 10^2$$

$$x^2 + x^2 + 4x + 4 = 100$$

$$2x^2 + 4x + 4 = 100$$

$$2x^2 + 4x - 96 = 0$$

$$2(x^2 + 2x - 48) = 0$$

$$2(x+8)(x-6) = 0$$

$$x = 6 \quad (x = -8 \text{ is impossible})$$

$$x + 2 = 8$$

The legs are 6 inches and 8 inches.

19. Let x represent the longer leg and $x + 1$ represent the hypotenuse. The equation is:

$$5^2 + x^2 = (x+1)^2$$

$$25 + x^2 = x^2 + 2x + 1$$

$$25 = 2x + 1$$

$$24 = 2x$$

$$x = 12$$

The longer leg is 12 meters.

21. Setting $C = \$1,400$:

$$1400 = 400 + 700x - 100x^2$$

$$100x^2 - 700x + 1000 = 0$$

$$100(x^2 - 7x + 10) = 0$$

$$100(x-5)(x-2) = 0$$

$$x = 2, 5$$

The company can manufacture either 200 items or 500 items.

23.   The revenue is given by: $R = xp = (1700 - 100p)p$. Setting $R = \$7,000$:

$$7000 = (1700 - 100p)p$$
$$7000 = 1700p - 100p^2$$
$$100p^2 - 1700p + 7000 = 0$$
$$100(p^2 - 17p + 70) = 0$$
$$100(p - 7)(p - 10) = 0$$
$$p = 7, 10$$

The calculators should be sold for either $7 or $10.

25.   **a.**   Let x represent the distance from the base to the wall, and $2x + 2$ represent the height on the wall. Using the Pythagorean theorem:

$$x^2 + (2x + 2)^2 = 13^2$$
$$x^2 + 4x^2 + 8x + 4 = 169$$
$$5x^2 + 8x - 165 = 0$$
$$(5x + 33)(x - 5) = 0$$
$$x = 5 \quad \left(x = -\frac{33}{5} \text{ is impossible} \right)$$

The base of the ladder is 5 feet from the wall.

   b.   Since $2x + 2 = 2 \cdot 5 + 2 = 12$, the ladder reaches a height of 12 feet.

27.   **a.**   Finding when $h = 0$:

$$0 = -16t^2 + 396t + 100$$
$$16t^2 - 396t - 100 = 0$$
$$4t^2 - 99t - 25 = 0$$
$$(4t + 1)(t - 25) = 0$$
$$t = 25 \quad \left(t = -\frac{1}{4} \text{ is impossible} \right)$$

The bullet will land on the ground after 25 seconds.

   b.   Completing the table:

t (seconds)	h (feet)
0	100
5	1,680
10	2,460
15	2,440
20	1,620
25	0

29.   Simplifying the expression: $(5x^3)^2 (2x^6)^3 = 25x^6 \cdot 8x^{18} = 200x^{24}$

31.   Simplifying the expression: $\dfrac{x^4}{x^{-3}} = x^{4-(-3)} = x^{4+3} = x^7$

33. Simplifying the expression: $\left(2\times10^{-4}\right)\left(4\times10^{5}\right)=8\times10^{1}$

35. Simplifying the expression: $20ab^{2}-16ab^{2}+6ab^{2}=10ab^{2}$

37. Multiplying using the distributive property:
$$2x^{2}\left(3x^{2}+3x-1\right)=2x^{2}\left(3x^{2}\right)+2x^{2}\left(3x\right)-2x^{2}\left(1\right)=6x^{4}+6x^{3}-2x^{2}$$

39. Multiplying using the square of binomial formula:
$$\left(3y-5\right)^{2}=\left(3y\right)^{2}-2\left(3y\right)\left(5\right)+\left(5\right)^{2}=9y^{2}-30y+25$$

41. Multiplying using the difference of squares formula:
$$\left(2a^{2}+7\right)\left(2a^{2}-7\right)=\left(2a^{2}\right)^{2}-\left(7\right)^{2}=4a^{4}-49$$

Chapter 5 Test

See www.mathtv.com for video solutions to all problems in this chapter test.

Chapter 6
Rational Expressions

6.1 Reducing Rational Expressions to Lowest Terms

1. Reducing the rational expression: $\dfrac{5}{5x-10} = \dfrac{5}{5(x-2)} = \dfrac{1}{x-2}$.

   The variable restriction is $x \neq 2$.

3. Reducing the rational expression: $\dfrac{a-3}{a^2-9} = \dfrac{1(a-3)}{(a+3)(a-3)} = \dfrac{1}{a+3}$.

   The variable restriction is $a \neq -3, 3$.

5. Reducing the rational expression: $\dfrac{x+5}{x^2-25} = \dfrac{1(x+5)}{(x+5)(x-5)} = \dfrac{1}{x-5}$.

   The variable restriction is $x \neq -5, 5$.

7. Reducing the rational expression: $\dfrac{2x^2-8}{4} = \dfrac{2(x^2-4)}{4} = \dfrac{2(x+2)(x-2)}{4} = \dfrac{(x+2)(x-2)}{2}$

   There are no variable restrictions.

9. Reducing the rational expression: $\dfrac{2x-10}{3x-6} = \dfrac{2(x-5)}{3(x-2)}$.

   The variable restriction is $x \neq 2$.

11. Reducing the rational expression: $\dfrac{10a+20}{5a+10} = \dfrac{10(a+2)}{5(a+2)} = \dfrac{2}{1} = 2$

13. Reducing the rational expression: $\dfrac{5x^2-5}{4x+4} = \dfrac{5(x^2-1)}{4(x+1)} = \dfrac{5(x+1)(x-1)}{4(x+1)} = \dfrac{5(x-1)}{4}$

15. Reducing the rational expression: $\dfrac{x-3}{x^2-6x+9} = \dfrac{1(x-3)}{(x-3)^2} = \dfrac{1}{x-3}$

17. Reducing the rational expression: $\dfrac{3x+15}{3x^2+24x+45} = \dfrac{3(x+5)}{3(x^2+8x+15)} = \dfrac{3(x+5)}{3(x+5)(x+3)} = \dfrac{1}{x+3}$

19. Reducing the rational expression: $\dfrac{a^2-3a}{a^3-8a^2+15a} = \dfrac{a(a-3)}{a(a^2-8a+15)} = \dfrac{a(a-3)}{a(a-3)(a-5)} = \dfrac{1}{a-5}$

21. Reducing the rational expression: $\dfrac{3x-2}{9x^2-4} = \dfrac{1(3x-2)}{(3x+2)(3x-2)} = \dfrac{1}{3x+2}$

23. Reducing the rational expression: $\dfrac{x^2+8x+15}{x^2+5x+6} = \dfrac{(x+5)(x+3)}{(x+2)(x+3)} = \dfrac{x+5}{x+2}$

25. Reducing the rational expression:

$$\frac{2m^3 - 2m^2 - 12m}{m^2 - 5m + 6} = \frac{2m(m^2 - m - 6)}{(m-2)(m-3)} = \frac{2m(m-3)(m+2)}{(m-2)(m-3)} = \frac{2m(m+2)}{m-2}$$

27. Reducing the rational expression: $\dfrac{x^3 + 3x^2 - 4x}{x^3 - 16x} = \dfrac{x(x^2 + 3x - 4)}{x(x^2 - 16)} = \dfrac{x(x+4)(x-1)}{x(x+4)(x-4)} = \dfrac{x-1}{x-4}$

29. Reducing the rational expression:

$$\frac{4x^3 - 10x^2 + 6x}{2x^3 + x^2 - 3x} = \frac{2x(2x^2 - 5x + 3)}{x(2x^2 + x - 3)} = \frac{2x(2x-3)(x-1)}{x(2x+3)(x-1)} = \frac{2(2x-3)}{2x+3}$$

31. Reducing the rational expression: $\dfrac{4x^2 - 12x + 9}{4x^2 - 9} = \dfrac{(2x-3)^2}{(2x+3)(2x-3)} = \dfrac{2x-3}{2x+3}$

33. Reducing the rational expression:

$$\frac{x+3}{x^4 - 81} = \frac{x+3}{(x^2+9)(x^2-9)} = \frac{x+3}{(x^2+9)(x+3)(x-3)} = \frac{1}{(x^2+9)(x-3)}$$

35. Reducing the rational expression:

$$\frac{3x^2 + x - 10}{x^4 - 16} = \frac{(3x-5)(x+2)}{(x^2+4)(x^2-4)} = \frac{(3x-5)(x+2)}{(x^2+4)(x+2)(x-2)} = \frac{3x-5}{(x^2+4)(x-2)}$$

37. Reducing the rational expression:

$$\frac{42x^3 - 20x^2 - 48x}{6x^2 - 5x - 4} = \frac{2x(21x^2 - 10x - 24)}{(3x-4)(2x+1)} = \frac{2x(7x+6)(3x-4)}{(3x-4)(2x+1)} = \frac{2x(7x+6)}{2x+1}$$

39. Reducing the rational expression: $\dfrac{x^3 - y^3}{x^2 - y^2} = \dfrac{(x-y)(x^2 + xy + y^2)}{(x+y)(x-y)} = \dfrac{x^2 + xy + y^2}{x+y}$

41. Reducing the rational expression: $\dfrac{x^3 + 8}{x^2 - 4} = \dfrac{(x+2)(x^2 - 2x + 4)}{(x+2)(x-2)} = \dfrac{x^2 - 2x + 4}{x-2}$

43. Reducing the rational expression: $\dfrac{x^3 + 8}{x^2 + x - 2} = \dfrac{(x+2)(x^2 - 2x + 4)}{(x+2)(x-1)} = \dfrac{x^2 - 2x + 4}{x-1}$

45. Reducing the rational expression: $\dfrac{xy + 3x + 2y + 6}{xy + 3x + 5y + 15} = \dfrac{x(y+3) + 2(y+3)}{x(y+3) + 5(y+3)} = \dfrac{(y+3)(x+2)}{(y+3)(x+5)} = \dfrac{x+2}{x+5}$

47. Reducing the rational expression: $\dfrac{x^2 - 3x + ax - 3a}{x^2 - 3x + bx - 3b} = \dfrac{x(x-3) + a(x-3)}{x(x-3) + b(x-3)} = \dfrac{(x-3)(x+a)}{(x-3)(x+b)} = \dfrac{x+a}{x+b}$

49. **a.** Adding: $(x^2 - 4x) + (4x - 16) = x^2 - 4x + 4x - 16 = x^2 - 16$

   b. Subtracting: $(x^2 - 4x) - (4x - 16) = x^2 - 4x - 4x + 16 = x^2 - 8x + 16$

   c. Multiplying: $(x^2 - 4x)(4x - 16) = 4x^3 - 16x^2 - 16x^2 + 64x = 4x^3 - 32x^2 + 64x$

   d. Reducing: $\dfrac{x^2 - 4x}{4x - 16} = \dfrac{x(x-4)}{4(x-4)} = \dfrac{x}{4}$

51. Writing as a ratio: $\dfrac{8}{6} = \dfrac{4}{3}$

53. Writing as a ratio: $\dfrac{200}{250} = \dfrac{4}{5}$

55. Writing as a ratio: $\dfrac{32}{4} = \dfrac{8}{1}$

57. Completing the table:

Checks Written x	Total Cost $2.00 + 0.15x$	Cost per Check $\dfrac{2.00 + 0.15x}{x}$
0	$2.00	undefined
5	$2.75	$0.55
10	$3.50	$0.35
15	$4.25	$0.28
20	$5.00	$0.25

59. The average speed is: $\dfrac{122 \text{ miles}}{3 \text{ hours}} \approx 40.7$ miles/hour

61. Computing the strikeouts per inning:

Randy Johnson: $\dfrac{10.95 \text{ strikeouts}}{9 \text{ innings}} \approx 1.22$ strikeouts/inning

Kerry Wood: $\dfrac{10.44 \text{ strikeouts}}{9 \text{ innings}} = 1.16$ strikeouts/inning

Pedro Martinez: $\dfrac{10.25 \text{ strikeouts}}{9 \text{ innings}} \approx 1.14$ strikeouts/inning

Curt Schilling: $\dfrac{8.77 \text{ strikeouts}}{9 \text{ innings}} \approx 0.97$ strikeouts/inning

63. The average speed is: $\dfrac{3 \text{ miles}}{24 \text{ minutes}} = 0.125$ miles/minute

65. Simplifying: $\dfrac{3}{4} \cdot \dfrac{10}{21} = \dfrac{3 \cdot 2 \cdot 5}{2 \cdot 2 \cdot 3 \cdot 7} = \dfrac{5}{14}$

67. Simplifying: $\dfrac{4}{5} \div \dfrac{8}{9} = \dfrac{4}{5} \cdot \dfrac{9}{8} = \dfrac{2 \cdot 2 \cdot 9}{5 \cdot 2 \cdot 2 \cdot 2} = \dfrac{9}{10}$

69. Factoring: $x^2 - 9 = (x+3)(x-3)$

71. Factoring: $3x - 9 = 3(x-3)$

73. Factoring: $x^2 - x - 20 = (x-5)(x+4)$

75. Factoring: $a^2 + 5a = a(a+5)$

77. Simplifying: $\dfrac{a(a+5)(a-5)(a+4)}{a^2 + 5a} = \dfrac{a(a+5)(a-5)(a+4)}{a(a+5)} = (a-5)(a+4)$

79. Multiplying: $\dfrac{5{,}603}{11} \cdot \dfrac{1}{2{,}580} \cdot \dfrac{60}{1} \approx 11.8$

6.2 Multiplication and Division of Rational Expressions

1.   Simplifying the expression: $\dfrac{x+y}{3} \cdot \dfrac{6}{x+y} = \dfrac{6(x+y)}{3(x+y)} = 2$

3.   Simplifying the expression: $\dfrac{2x+10}{x^2} \cdot \dfrac{x^3}{4x+20} = \dfrac{2(x+5)}{x^2} \cdot \dfrac{x^3}{4(x+5)} = \dfrac{2x^3(x+5)}{4x^2(x+5)} = \dfrac{x}{2}$

5.   Simplifying the expression: $\dfrac{9}{2a-8} \div \dfrac{3}{a-4} = \dfrac{9}{2a-8} \cdot \dfrac{a-4}{3} = \dfrac{9}{2(a-4)} \cdot \dfrac{a-4}{3} = \dfrac{9(a-4)}{6(a-4)} = \dfrac{3}{2}$

7.   Simplifying the expression:

$$\dfrac{x+1}{x^2-9} \div \dfrac{2x+2}{x+3} = \dfrac{x+1}{x^2-9} \cdot \dfrac{x+3}{2x+2}$$

$$= \dfrac{x+1}{(x+3)(x-3)} \cdot \dfrac{x+3}{2(x+1)}$$

$$= \dfrac{(x+1)(x+3)}{2(x+3)(x-3)(x+1)}$$

$$= \dfrac{1}{2(x-3)}$$

9.   Simplifying the expression: $\dfrac{a^2+5a}{7a} \cdot \dfrac{4a^2}{a^2+4a} = \dfrac{a(a+5)}{7a} \cdot \dfrac{4a^2}{a(a+4)} = \dfrac{4a^3(a+5)}{7a^2(a+4)} = \dfrac{4a(a+5)}{7(a+4)}$

11.   Simplifying the expression:

$$\dfrac{y^2-5y+6}{2y+4} \div \dfrac{2y-6}{y+2} = \dfrac{y^2-5y+6}{2y+4} \cdot \dfrac{y+2}{2y-6}$$

$$= \dfrac{(y-2)(y-3)}{2(y+2)} \cdot \dfrac{y+2}{2(y-3)}$$

$$= \dfrac{(y-2)(y-3)(y+2)}{4(y+2)(y-3)}$$

$$= \dfrac{y-2}{4}$$

13.   Simplifying the expression:

$$\dfrac{2x-8}{x^2-4} \cdot \dfrac{x^2+6x+8}{x-4} = \dfrac{2(x-4)}{(x+2)(x-2)} \cdot \dfrac{(x+4)(x+2)}{x-4} = \dfrac{2(x-4)(x+4)(x+2)}{(x+2)(x-2)(x-4)} = \dfrac{2(x+4)}{x-2}$$

15. Simplifying the expression:

$$\frac{x-1}{x^2-x-6} \cdot \frac{x^2+5x+6}{x^2-1} = \frac{x-1}{(x-3)(x+2)} \cdot \frac{(x+2)(x+3)}{(x+1)(x-1)}$$

$$= \frac{(x-1)(x+2)(x+3)}{(x-3)(x+2)(x+1)(x-1)}$$

$$= \frac{x+3}{(x-3)(x+1)}$$

17. Simplifying the expression:

$$\frac{a^2+10a+25}{a+5} \div \frac{a^2-25}{a-5} = \frac{a^2+10a+25}{a+5} \cdot \frac{a-5}{a^2-25}$$

$$= \frac{(a+5)^2}{a+5} \cdot \frac{a-5}{(a+5)(a-5)}$$

$$= \frac{(a+5)^2(a-5)}{(a+5)^2(a-5)}$$

$$= 1$$

19. Simplifying the expression:

$$\frac{y^3-5y^2}{y^4+3y^3+2y^2} \div \frac{y^2-5y+6}{y^2-2y-3} = \frac{y^3-5y^2}{y^4+3y^3+2y^2} \cdot \frac{y^2-2y-3}{y^2-5y+6}$$

$$= \frac{y^2(y-5)}{y^2(y+2)(y+1)} \cdot \frac{(y-3)(y+1)}{(y-2)(y-3)}$$

$$= \frac{y^2(y-5)(y-3)(y+1)}{y^2(y+2)(y+1)(y-2)(y-3)}$$

$$= \frac{y-5}{(y+2)(y-2)}$$

21. Simplifying the expression:

$$\frac{2x^2+17x+21}{x^2+2x-35} \cdot \frac{x^2-25}{2x^2-7x-15} = \frac{(2x+3)(x+7)}{(x+7)(x-5)} \cdot \frac{(x+5)(x-5)}{(2x+3)(x-5)}$$

$$= \frac{(2x+3)(x+7)(x+5)(x-5)}{(x+7)(x-5)^2(2x+3)}$$

$$= \frac{x+5}{x-5}$$

23. Simplifying the expression:

$$\frac{2x^2+10x+12}{4x^2+24x+32} \cdot \frac{2x^2+18x+40}{x^2+8x+15} = \frac{2\left(x^2+5x+6\right)}{4\left(x^2+6x+8\right)} \cdot \frac{2\left(x^2+9x+20\right)}{x^2+8x+15}$$

$$= \frac{2(x+2)(x+3)}{4(x+4)(x+2)} \cdot \frac{2(x+5)(x+4)}{(x+5)(x+3)}$$

$$= \frac{4(x+2)(x+3)(x+4)(x+5)}{4(x+2)(x+3)(x+4)(x+5)}$$

$$= 1$$

25. Simplifying the expression:

$$\frac{2a^2+7a+3}{a^2-16} \div \frac{4a^2+8a+3}{2a^2-5a-12} = \frac{2a^2+7a+3}{a^2-16} \cdot \frac{2a^2-5a-12}{4a^2+8a+3}$$

$$= \frac{(2a+1)(a+3)}{(a+4)(a-4)} \cdot \frac{(2a+3)(a-4)}{(2a+1)(2a+3)}$$

$$= \frac{(2a+1)(a+3)(2a+3)(a-4)}{(a+4)(a-4)(2a+1)(2a+3)}$$

$$= \frac{a+3}{a+4}$$

27. Simplifying the expression:

$$\frac{4y^2-12y+9}{y^2-36} \div \frac{2y^2-5y+3}{y^2+5y-6} = \frac{4y^2-12y+9}{y^2-36} \cdot \frac{y^2+5y-6}{2y^2-5y+3}$$

$$= \frac{(2y-3)^2}{(y+6)(y-6)} \cdot \frac{(y+6)(y-1)}{(2y-3)(y-1)}$$

$$= \frac{(2y-3)^2(y+6)(y-1)}{(y+6)(y-6)(2y-3)(y-1)}$$

$$= \frac{2y-3}{y-6}$$

29. Simplifying the expression:

$$\frac{x^2-1}{6x^2+18x+12} \cdot \frac{7x^2+17x+6}{x+1} \cdot \frac{6x+30}{7x^2-11x-6}$$

$$= \frac{(x+1)(x-1)}{6(x+2)(x+1)} \cdot \frac{(7x+3)(x+2)}{x+1} \cdot \frac{6(x+5)}{(7x+3)(x-2)}$$

$$= \frac{6(x+1)(x-1)(7x+3)(x+2)(x+5)}{6(x+2)(x+1)(x+1)(7x+3)(x-2)}$$

$$= \frac{(x-1)(x+5)}{(x+1)(x-2)}$$

31. Simplifying the expression:

$$\frac{18x^3 + 21x^2 - 60x}{21x^2 - 25x - 4} \cdot \frac{28x^2 - 17x - 3}{16x^3 + 28x^2 - 30x} = \frac{3x\left(6x^2 + 7x - 20\right)}{21x^2 - 25x - 4} \cdot \frac{28x^2 - 17x - 3}{2x\left(8x^2 + 14x - 15\right)}$$

$$= \frac{3x(3x - 4)(2x + 5)}{(7x + 1)(3x - 4)} \cdot \frac{(7x + 1)(4x - 3)}{2x(4x - 3)(2x + 5)}$$

$$= \frac{3x(3x - 4)(2x + 5)(7x + 1)(4x - 3)}{2x(7x + 1)(3x - 4)(4x - 3)(2x + 5)}$$

$$= \frac{3}{2}$$

33. **a.** Simplifying: $\dfrac{9 - 1}{27 - 1} = \dfrac{8}{26} = \dfrac{4}{13}$

   b. Reducing: $\dfrac{x^2 - 1}{x^3 - 1} = \dfrac{(x + 1)(x - 1)}{(x - 1)(x^2 + x + 1)} = \dfrac{x + 1}{x^2 + x + 1}$

   c. Multiplying: $\dfrac{x^2 - 1}{x^3 - 1} \cdot \dfrac{x - 2}{x + 1} = \dfrac{(x + 1)(x - 1)}{(x - 1)(x^2 + x + 1)} \cdot \dfrac{x - 2}{x + 1} = \dfrac{x - 2}{x^2 + x + 1}$

   d. Dividing: $\dfrac{x^2 - 1}{x^3 - 1} \div \dfrac{x - 1}{x^2 + x + 1} = \dfrac{(x + 1)(x - 1)}{(x - 1)(x^2 + x + 1)} \cdot \dfrac{x^2 + x + 1}{x - 1} = \dfrac{x + 1}{x - 1}$

35. Simplifying the expression: $\left(x^2 - 9\right)\left(\dfrac{2}{x + 3}\right) = \dfrac{(x + 3)(x - 3)}{1} \cdot \dfrac{2}{x + 3} = \dfrac{2(x + 3)(x - 3)}{x + 3} = 2(x - 3)$

37. Simplifying the expression:

$$\left(x^2 - x - 6\right)\left(\frac{x + 1}{x - 3}\right) = \frac{(x - 3)(x + 2)}{1} \cdot \frac{x + 1}{x - 3} = \frac{(x - 3)(x + 2)(x + 1)}{x - 3} = (x + 2)(x + 1)$$

39. Simplifying the expression:

$$\left(x^2 - 4x - 5\right)\left(\frac{-2x}{x + 1}\right) = \frac{(x - 5)(x + 1)}{1} \cdot \frac{-2x}{x + 1} = \frac{-2x(x - 5)(x + 1)}{x + 1} = -2x(x - 5)$$

41. Simplifying the expression:

$$\frac{x^2 - 9}{x^2 - 3x} \cdot \frac{2x + 10}{xy + 5x + 3y + 15} = \frac{(x + 3)(x - 3)}{x(x - 3)} \cdot \frac{2(x + 5)}{x(y + 5) + 3(y + 5)}$$

$$= \frac{2(x + 3)(x - 3)(x + 5)}{x(x - 3)(y + 5)(x + 3)}$$

$$= \frac{2(x + 5)}{x(y + 5)}$$

43. Simplifying the expression:

$$\frac{2x^2+4x}{x^2-y^2}\cdot\frac{x^2+3x+xy+3y}{x^2+5x+6}=\frac{2x(x+2)}{(x+y)(x-y)}\cdot\frac{x(x+3)+y(x+3)}{(x+2)(x+3)}$$

$$=\frac{2x(x+2)(x+3)(x+y)}{(x+y)(x-y)(x+2)(x+3)}$$

$$=\frac{2x}{x-y}$$

45. Simplifying the expression:

$$\frac{x^3-3x^2+4x-12}{x^4-16}\cdot\frac{3x^2+5x-2}{3x^2-10x+3}=\frac{x^2(x-3)+4(x-3)}{(x^2+4)(x^2-4)}\cdot\frac{(3x-1)(x+2)}{(3x-1)(x-3)}$$

$$=\frac{(x-3)(x^2+4)}{(x^2+4)(x+2)(x-2)}\cdot\frac{(3x-1)(x+2)}{(3x-1)(x-3)}$$

$$=\frac{(x-3)(x^2+4)(3x-1)(x+2)}{(x^2+4)(x+2)(x-2)(3x-1)(x-3)}$$

$$=\frac{1}{x-2}$$

47. Simplifying the expression:

$$\left(1-\frac{1}{2}\right)\left(1-\frac{1}{3}\right)\left(1-\frac{1}{4}\right)\left(1-\frac{1}{5}\right)=\left(\frac{2}{2}-\frac{1}{2}\right)\left(\frac{3}{3}-\frac{1}{3}\right)\left(\frac{4}{4}-\frac{1}{4}\right)\left(\frac{5}{5}-\frac{1}{5}\right)=\frac{1}{2}\cdot\frac{2}{3}\cdot\frac{3}{4}\cdot\frac{4}{5}=\frac{1}{5}$$

49. Simplifying the expression:

$$\left(1-\frac{1}{2}\right)\left(1-\frac{1}{3}\right)\left(1-\frac{1}{4}\right)\cdots\left(1-\frac{1}{99}\right)\left(1-\frac{1}{100}\right)=\frac{1}{2}\cdot\frac{2}{3}\cdot\frac{3}{4}\cdots\frac{98}{99}\cdot\frac{99}{100}=\frac{1}{100}$$

51. Since 5,280 feet = 1 mile, the height is: $\dfrac{14,494\text{ feet}}{5,280\text{ feet/mile}}\approx 2.7$ miles

53. Converting to miles per hour: $\dfrac{1088\text{ feet}}{1\text{ second}}\cdot\dfrac{1\text{ mile}}{5280\text{ feet}}\cdot\dfrac{60\text{ seconds}}{1\text{ minute}}\cdot\dfrac{60\text{ minutes}}{1\text{ hour}}\approx 742$ miles/hour

55. Converting to miles per hour: $\dfrac{785\text{ feet}}{20\text{ minutes}}\cdot\dfrac{60\text{ minutes}}{1\text{ hour}}\cdot\dfrac{1\text{ mile}}{5280\text{ feet}}\approx 0.45$ miles/hour

57. Converting to miles per hour: $\dfrac{518\text{ feet}}{40\text{ seconds}}\cdot\dfrac{60\text{ seconds}}{1\text{ minute}}\cdot\dfrac{60\text{ minutes}}{1\text{ hour}}\cdot\dfrac{1\text{ mile}}{5280\text{ feet}}\approx 8.8$ miles/hour

59. Adding the fractions: $\dfrac{1}{5}+\dfrac{3}{5}=\dfrac{4}{5}$

61. Adding the fractions: $\dfrac{1}{10}+\dfrac{3}{14}=\dfrac{1}{10}\cdot\dfrac{7}{7}+\dfrac{3}{14}\cdot\dfrac{5}{5}=\dfrac{7}{70}+\dfrac{15}{70}=\dfrac{22}{70}=\dfrac{11}{35}$

63. Subtracting the fractions: $\dfrac{1}{10}-\dfrac{3}{14}=\dfrac{1}{10}\cdot\dfrac{7}{7}-\dfrac{3}{14}\cdot\dfrac{5}{5}=\dfrac{7}{70}-\dfrac{15}{70}=-\dfrac{8}{70}=-\dfrac{4}{35}$

65. Multiplying: $2(x-3)=2x-6$

67. Multiplying: $(x+4)(x-5) = x^2 - 5x + 4x - 20 = x^2 - x - 20$

69. Reducing the fraction: $\dfrac{x+3}{x^2-9} = \dfrac{x+3}{(x+3)(x-3)} = \dfrac{1}{x-3}$

71. Reducing the fraction: $\dfrac{x^2-x-30}{2(x+5)(x-5)} = \dfrac{(x+5)(x-6)}{2(x+5)(x-5)} = \dfrac{x-6}{2(x-5)}$

73. Simplifying: $(x+4)(x-5) - 10 = x^2 - 5x + 4x - 20 - 10 = x^2 - x - 30$

6.3 Addition and Subtraction of Rational Expressions

1. Combining the fractions: $\dfrac{3}{x} + \dfrac{4}{x} = \dfrac{7}{x}$ **3.** Combining the fractions: $\dfrac{9}{a} - \dfrac{5}{a} = \dfrac{4}{a}$

5. Combining the fractions: $\dfrac{1}{x+1} + \dfrac{x}{x+1} = \dfrac{1+x}{x+1} = 1$

7. Combining the fractions: $\dfrac{y^2}{y-1} - \dfrac{1}{y-1} = \dfrac{y^2-1}{y-1} = \dfrac{(y+1)(y-1)}{y-1} = y+1$

9. Combining the fractions: $\dfrac{x^2}{x+2} + \dfrac{4x+4}{x+2} = \dfrac{x^2+4x+4}{x+2} = \dfrac{(x+2)^2}{x+2} = x+2$

11. Combining the fractions: $\dfrac{x^2}{x-2} - \dfrac{4x-4}{x-2} = \dfrac{x^2-4x+4}{x-2} = \dfrac{(x-2)^2}{x-2} = x-2$

13. Combining the fractions: $\dfrac{x+2}{x+6} - \dfrac{x-4}{x+6} = \dfrac{x+2-x+4}{x+6} = \dfrac{6}{x+6}$

15. Combining the fractions: $\dfrac{y}{2} - \dfrac{2}{y} = \dfrac{y \cdot y}{2 \cdot y} - \dfrac{2 \cdot 2}{y \cdot 2} = \dfrac{y^2}{2y} - \dfrac{4}{2y} = \dfrac{y^2-4}{2y} = \dfrac{(y+2)(y-2)}{2y}$

17. Combining the fractions: $\dfrac{1}{2} + \dfrac{a}{3} = \dfrac{1 \cdot 3}{2 \cdot 3} + \dfrac{a \cdot 2}{3 \cdot 2} = \dfrac{3}{6} + \dfrac{2a}{6} = \dfrac{2a+3}{6}$

19. Combining the fractions:

$$\dfrac{x}{x+1} + \dfrac{3}{4} = \dfrac{x \cdot 4}{(x+1) \cdot 4} + \dfrac{3 \cdot (x+1)}{4 \cdot (x+1)} = \dfrac{4x}{4(x+1)} + \dfrac{3x+3}{4(x+1)} = \dfrac{4x+3x+3}{4(x+1)} = \dfrac{7x+3}{4(x+1)}$$

21. Combining the fractions:

$$\dfrac{x+1}{x-2} - \dfrac{4x+7}{5x-10} = \dfrac{(x+1) \cdot 5}{(x-2) \cdot 5} - \dfrac{4x+7}{5(x-2)}$$

$$= \dfrac{5x+5}{5(x-2)} - \dfrac{4x+7}{5(x-2)}$$

$$= \dfrac{5x+5-4x-7}{5(x-2)}$$

$$= \dfrac{x-2}{5(x-2)}$$

$$= \dfrac{1}{5}$$

23.   Combining the fractions:

$$\frac{4x-2}{3x+12}-\frac{x-2}{x+4}=\frac{4x-2}{3(x+4)}-\frac{(x-2)\bullet 3}{(x+4)\bullet 3}$$

$$=\frac{4x-2}{3(x+4)}-\frac{3x-6}{3(x+4)}$$

$$=\frac{4x-2-3x+6}{3(x+4)}$$

$$=\frac{x+4}{3(x+4)}$$

$$=\frac{1}{3}$$

25.   Combining the fractions:

$$\frac{6}{x(x-2)}+\frac{3}{x}=\frac{6}{x(x-2)}+\frac{3\bullet(x-2)}{x\bullet(x-2)}=\frac{6}{x(x-2)}+\frac{3x-6}{x(x-2)}=\frac{6+3x-6}{x(x-2)}=\frac{3x}{x(x-2)}=\frac{3}{x-2}$$

27.   Combining the fractions:

$$\frac{4}{a}-\frac{12}{a^2+3a}=\frac{4\bullet(a+3)}{a\bullet(a+3)}-\frac{12}{a(a+3)}=\frac{4a+12}{a(a+3)}-\frac{12}{a(a+3)}=\frac{4a+12-12}{a(a+3)}=\frac{4a}{a(a+3)}=\frac{4}{a+3}$$

29.   Combining the fractions:

$$\frac{2}{x+5}-\frac{10}{x^2-25}=\frac{2\bullet(x-5)}{(x+5)\bullet(x-5)}-\frac{10}{(x+5)(x-5)}$$

$$=\frac{2x-10}{(x+5)(x-5)}-\frac{10}{(x+5)(x-5)}$$

$$=\frac{2x-10-10}{(x+5)(x-5)}$$

$$=\frac{2x-20}{(x+5)(x-5)}$$

$$=\frac{2(x-10)}{(x+5)(x-5)}$$

31.   Combining the fractions:

$$\frac{x-4}{x-3}+\frac{6}{x^2-9}=\frac{(x-4)\bullet(x+3)}{(x-3)\bullet(x+3)}+\frac{6}{(x+3)(x-3)}$$

$$=\frac{x^2-x-12}{(x+3)(x-3)}+\frac{6}{(x+3)(x-3)}$$

$$=\frac{x^2-x-12+6}{(x+3)(x-3)}$$

$$=\frac{x^2-x-6}{(x+3)(x-3)}$$

$$=\frac{(x-3)(x+2)}{(x+3)(x-3)}$$

$$=\frac{x+2}{x+3}$$

33. Combining the fractions:

$$\frac{a-4}{a-3}+\frac{5}{a^2-a-6}=\frac{(a-4)\bullet(a+2)}{(a-3)\bullet(a+2)}+\frac{5}{(a-3)(a+2)}$$

$$=\frac{a^2-2a-8}{(a-3)(a+2)}+\frac{5}{(a-3)(a+2)}$$

$$=\frac{a^2-2a-8+5}{(a-3)(a+2)}$$

$$=\frac{a^2-2a-3}{(a-3)(a+2)}$$

$$=\frac{(a-3)(a+1)}{(a-3)(a+2)}$$

$$=\frac{a+1}{a+2}$$

35. Combining the fractions:

$$\frac{8}{x^2-16}-\frac{7}{x^2-x-12}=\frac{8}{(x+4)(x-4)}-\frac{7}{(x-4)(x+3)}$$

$$=\frac{8(x+3)}{(x+4)(x-4)(x+3)}-\frac{7(x+4)}{(x+4)(x-4)(x+3)}$$

$$=\frac{8x+24}{(x+4)(x-4)(x+3)}-\frac{7x+28}{(x+4)(x-4)(x+3)}$$

$$=\frac{8x+24-7x-28}{(x+4)(x-4)(x+3)}$$

$$=\frac{x-4}{(x+4)(x-4)(x+3)}$$

$$=\frac{1}{(x+4)(x+3)}$$

37. Combining the fractions:

$$\frac{4y}{y^2+6y+5}-\frac{3y}{y^2+5y+4}=\frac{4y}{(y+5)(y+1)}-\frac{3y}{(y+4)(y+1)}$$

$$=\frac{4y(y+4)}{(y+5)(y+1)(y+4)}-\frac{3y(y+5)}{(y+5)(y+1)(y+4)}$$

$$=\frac{4y^2+16y}{(y+5)(y+1)(y+4)}-\frac{3y^2+15y}{(y+5)(y+1)(y+4)}$$

$$=\frac{4y^2+16y-3y^2-15y}{(y+5)(y+1)(y+4)}$$

$$=\frac{y^2+y}{(y+5)(y+1)(y+4)}$$

$$=\frac{y(y+1)}{(y+5)(y+1)(y+4)}$$

$$=\frac{y}{(y+5)(y+4)}$$

39. Combining the fractions:

$$\frac{4x+1}{x^2+5x+4}-\frac{x+3}{x^2+4x+3}=\frac{4x+1}{(x+4)(x+1)}-\frac{x+3}{(x+3)(x+1)}$$

$$=\frac{(4x+1)(x+3)}{(x+4)(x+1)(x+3)}-\frac{(x+3)(x+4)}{(x+4)(x+1)(x+3)}$$

$$=\frac{4x^2+13x+3}{(x+4)(x+1)(x+3)}-\frac{x^2+7x+12}{(x+4)(x+1)(x+3)}$$

$$=\frac{4x^2+13x+3-x^2-7x-12}{(x+4)(x+1)(x+3)}$$

$$=\frac{3x^2+6x-9}{(x+4)(x+1)(x+3)}$$

$$=\frac{3(x+3)(x-1)}{(x+4)(x+1)(x+3)}$$

$$=\frac{3(x-1)}{(x+4)(x+1)}$$

41. Combining the fractions:

$$\frac{1}{x}+\frac{x}{3x+9}-\frac{3}{x^2+3x}=\frac{1}{x}+\frac{x}{3(x+3)}-\frac{3}{x(x+3)}$$

$$=\frac{1\cdot 3(x+3)}{x\cdot 3(x+3)}+\frac{x\cdot x}{3(x+3)\cdot x}-\frac{3\cdot 3}{x(x+3)\cdot 3}$$

$$=\frac{3x+9}{3x(x+3)}+\frac{x^2}{3x(x+3)}-\frac{9}{3x(x+3)}$$

$$=\frac{3x+9+x^2-9}{3x(x+3)}$$

$$=\frac{x^2+3x}{3x(x+3)}$$

$$=\frac{x(x+3)}{3x(x+3)}$$

$$=\frac{1}{3}$$

43. a. Multiplying: $\dfrac{4}{9} \cdot \dfrac{1}{6} = \dfrac{2 \cdot 2}{3 \cdot 3} \cdot \dfrac{1}{2 \cdot 3} = \dfrac{2}{3 \cdot 3 \cdot 3} = \dfrac{2}{27}$

b. Dividing: $\dfrac{4}{9} \div \dfrac{1}{6} = \dfrac{4}{9} \cdot \dfrac{6}{1} = \dfrac{2 \cdot 2}{3 \cdot 3} \cdot \dfrac{2 \cdot 3}{1} = \dfrac{2 \cdot 2 \cdot 2}{3} = \dfrac{8}{3}$

c. Adding: $\dfrac{4}{9} + \dfrac{1}{6} = \dfrac{4}{9} \cdot \dfrac{2}{2} + \dfrac{1}{6} \cdot \dfrac{3}{3} = \dfrac{8}{18} + \dfrac{3}{18} = \dfrac{11}{18}$

d. Multiplying: $\dfrac{x+2}{x-2} \cdot \dfrac{3x+10}{x^2-4} = \dfrac{x+2}{x-2} \cdot \dfrac{3x+10}{(x+2)(x-2)} = \dfrac{3x+10}{(x-2)^2}$

e. Dividing: $\dfrac{x+2}{x-2} \div \dfrac{3x+10}{x^2-4} = \dfrac{x+2}{x-2} \cdot \dfrac{(x+2)(x-2)}{3x+10} = \dfrac{(x+2)^2}{3x+10}$

f. Subtracting:

$$\dfrac{x+2}{x-2} - \dfrac{3x+10}{x^2-4} = \dfrac{x+2}{x-2} \cdot \dfrac{x+2}{x+2} - \dfrac{3x+10}{(x+2)(x-2)}$$

$$= \dfrac{x^2+4x+4}{(x+2)(x-2)} - \dfrac{3x+10}{(x+2)(x-2)}$$

$$= \dfrac{x^2+x-6}{(x+2)(x-2)}$$

$$= \dfrac{(x+3)(x-2)}{(x+2)(x-2)}$$

$$= \dfrac{x+3}{x+2}$$

45. Completing the table:

Number x	Reciprocal $\dfrac{1}{x}$	Sum $1+\dfrac{1}{x}$	Sum $\dfrac{x+1}{x}$
1	1	2	2
2	$\frac{1}{2}$	$\frac{3}{2}$	$\frac{3}{2}$
3	$\frac{1}{3}$	$\frac{4}{3}$	$\frac{4}{3}$
4	$\frac{1}{4}$	$\frac{5}{4}$	$\frac{5}{4}$

47. Combining the fractions: $1 + \dfrac{1}{x+2} = \dfrac{1 \cdot (x+2)}{1 \cdot (x+2)} + \dfrac{1}{x+2} = \dfrac{x+2}{x+2} + \dfrac{1}{x+2} = \dfrac{x+2+1}{x+2} = \dfrac{x+3}{x+2}$

49. Combining the fractions: $1 - \dfrac{1}{x+3} = \dfrac{1 \cdot (x+3)}{1 \cdot (x+3)} - \dfrac{1}{x+3} = \dfrac{x+3}{x+3} - \dfrac{1}{x+3} = \dfrac{x+3-1}{x+3} = \dfrac{x+2}{x+3}$

51. The expression is: $x + 2\left(\dfrac{1}{x}\right) = x + \dfrac{2}{x} = \dfrac{x \cdot x}{1 \cdot x} + \dfrac{2}{x} = \dfrac{x^2}{x} + \dfrac{2}{x} = \dfrac{x^2+2}{x}$

53. Represent the two numbers as x and $2x$. The expression is: $\dfrac{1}{x} + \dfrac{1}{2x} = \dfrac{1 \cdot 2}{x \cdot 2} + \dfrac{1}{2x} = \dfrac{2}{2x} + \dfrac{1}{2x} = \dfrac{3}{2x}$

55. Simplifying: $6\left(\dfrac{1}{2}\right) = 3$

57. Simplifying: $\dfrac{0}{5} = 0$

59. Simplifying: $\dfrac{5}{0}$ is undefined

61. Simplifying: $1 - \dfrac{5}{2} = \dfrac{2}{2} - \dfrac{5}{2} = -\dfrac{3}{2}$

63. Simplifying: $6\left(\dfrac{x}{3} + \dfrac{5}{2}\right) = 6 \bullet \dfrac{x}{3} + 6 \bullet \dfrac{5}{2} = 2x + 15$

65. Simplifying: $x^2\left(1 - \dfrac{5}{x}\right) = x^2 \bullet 1 - x^2 \bullet \dfrac{5}{x} = x^2 - 5x$

67. Solving the equation:
$$2x + 15 = 3$$
$$2x = -12$$
$$x = -6$$

69. Solving the equation:
$$-2x - 9 = x - 3$$
$$-3x - 9 = -3$$
$$-3x = 6$$
$$x = -2$$

6.4 Equations Involving Rational Expressions

1. Multiplying both sides of the equation by 6:
$$6\left(\dfrac{x}{3} + \dfrac{1}{2}\right) = 6\left(-\dfrac{1}{2}\right)$$
$$2x + 3 = -3$$
$$2x = -6$$
$$x = -3$$
Since $x = -3$ checks in the original equation, the solution is $x = -3$.

3. Multiplying both sides of the equation by $5a$:
$$5a\left(\dfrac{4}{a}\right) = 5a\left(\dfrac{1}{5}\right)$$
$$20 = a$$
Since $a = 20$ checks in the original equation, the solution is $a = 20$.

5. Multiplying both sides of the equation by x:
$$x\left(\dfrac{3}{x} + 1\right) = x\left(\dfrac{2}{x}\right)$$
$$3 + x = 2$$
$$x = -1$$
Since $x = -1$ checks in the original equation, the solution is $x = -1$.

7. Multiplying both sides of the equation by $5a$:
$$5a\left(\dfrac{3}{a} - \dfrac{2}{a}\right) = 5a\left(\dfrac{1}{5}\right)$$
$$15 - 10 = a$$
$$a = 5$$
Since $a = 5$ checks in the original equation, the solution is $a = 5$.

9.   Multiplying both sides of the equation by $2x$:

$$2x\left(\frac{3}{x}+2\right)=2x\left(\frac{1}{2}\right)$$

$$6+4x=x$$

$$6=-3x$$

$$x=-2$$

Since $x=-2$ checks in the original equation, the solution is $x=-2$.

11.   Multiplying both sides of the equation by $4y$:

$$4y\left(\frac{1}{y}-\frac{1}{2}\right)=4y\left(-\frac{1}{4}\right)$$

$$4-2y=-y$$

$$4=y$$

Since $y=4$ checks in the original equation, the solution is $y=4$.

13.   Multiplying both sides of the equation by x^2:

$$x^2\left(1-\frac{8}{x}\right)=x^2\left(-\frac{15}{x^2}\right)$$

$$x^2-8x=-15$$

$$x^2-8x+15=0$$

$$(x-3)(x-5)=0$$

$$x=3,5$$

Both $x=3$ and $x=5$ check in the original equation.

15.   Multiplying both sides of the equation by $2x$:

$$2x\left(\frac{x}{2}-\frac{4}{x}\right)=2x\left(-\frac{7}{2}\right)$$

$$x^2-8=-7x$$

$$x^2+7x-8=0$$

$$(x+8)(x-1)=0$$

$$x=-8,1$$

Both $x=-8$ and $x=1$ check in the original equation.

17.   Multiplying both sides of the equation by 6:

$$6\left(\frac{x-3}{2}+\frac{2x}{3}\right)=6\left(\frac{5}{6}\right)$$

$$3(x-3)+2(2x)=5$$

$$3x-9+4x=5$$

$$7x-9=5$$

$$7x=14$$

$$x=2$$

Since $x=2$ checks in the original equation, the solution is $x=2$.

19. Multiplying both sides of the equation by 12:

$$12\left(\frac{x+1}{3}+\frac{x-3}{4}\right)=12\left(\tfrac{1}{6}\right)$$

$$4(x+1)+3(x-3)=2$$

$$4x+4+3x-9=2$$

$$7x-5=2$$

$$7x=7$$

$$x=1$$

Since $x=1$ checks in the original equation, the solution is $x=1$.

21. Multiplying both sides of the equation by $5(x+2)$:

$$5(x+2)\cdot\frac{6}{x+2}=5(x+2)\cdot\frac{3}{5}$$

$$30=3x+6$$

$$24=3x$$

$$x=8$$

Since $x=8$ checks in the original equation, the solution is $x=8$.

23. Multiplying both sides of the equation by $(y-2)(y-3)$:

$$(y-2)(y-3)\cdot\frac{3}{y-2}=(y-2)(y-3)\cdot\frac{2}{y-3}$$

$$3(y-3)=2(y-2)$$

$$3y-9=2y-4$$

$$y=5$$

Since $y=5$ checks in the original equation, the solution is $y=5$.

25. Multiplying both sides of the equation by $3(x-2)$:

$$3(x-2)\left(\frac{x}{x-2}+\frac{2}{3}\right)=3(x-2)\left(\frac{2}{x-2}\right)$$

$$3x+2(x-2)=6$$

$$3x+2x-4=6$$

$$5x-4=6$$

$$5x=10$$

$$x=2$$

Since $x=2$ does not check in the original equation, there is no solution ($\varnothing$).

27.   Multiplying both sides of the equation by $2(x-2)$:

$$2(x-2)\left(\frac{x}{x-2}+\frac{3}{2}\right)=2(x-2)\cdot\frac{9}{2(x-2)}$$

$$2x+3(x-2)=9$$

$$2x+3x-6=9$$

$$5x-6=9$$

$$5x=15$$

$$x=3$$

Since $x=3$ checks in the original equation, the solution is $x=3$.

29.   Multiplying both sides of the equation by $x^2+5x+6=(x+2)(x+3)$:

$$(x+2)(x+3)\left(\frac{5}{x+2}+\frac{1}{x+3}\right)=(x+2)(x+3)\cdot\frac{-1}{(x+2)(x+3)}$$

$$5(x+3)+1(x+2)=-1$$

$$5x+15+x+2=-1$$

$$6x+17=-1$$

$$6x=-18$$

$$x=-3$$

Since $x=-3$ does not check in the original equation, there is no solution ($\varnothing$).

31.   Multiplying both sides of the equation by $x^2-4=(x+2)(x-2)$:

$$(x+2)(x-2)\left(\frac{8}{x^2-4}+\frac{3}{x+2}\right)=(x+2)(x-2)\cdot\frac{1}{x-2}$$

$$8+3(x-2)=1(x+2)$$

$$8+3x-6=x+2$$

$$3x+2=x+2$$

$$2x=0$$

$$x=0$$

Since $x=0$ checks in the original equation, the solution is $x=0$.

33.   Multiplying both sides of the equation by $2(a-3)$:

$$2(a-3)\left(\frac{a}{2}+\frac{3}{a-3}\right)=2(a-3)\cdot\frac{a}{a-3}$$

$$a(a-3)+6=2a$$

$$a^2-3a+6=2a$$

$$a^2-5a+6=0$$

$$(a-2)(a-3)=0$$

$$a=2,3$$

Since $a=3$ does not check in the original equation, the solution is $a=2$.

35. Since $y^2 - 4 = (y+2)(y-2)$ and $y^2 + 2y = y(y+2)$, the LCD is $y(y+2)(y-2)$.
Multiplying by the LCD:

$$y(y+2)(y-2) \cdot \frac{6}{(y+2)(y-2)} = y(y+2)(y-2) \cdot \frac{4}{y(y+2)}$$

$$6y = 4(y-2)$$

$$6y = 4y - 8$$

$$2y = -8$$

$$y = -4$$

Since $y = -4$ checks in the original equation, the solution is $y = -4$.

37. Since $a^2 - 9 = (a+3)(a-3)$ and $a^2 + a - 12 = (a+4)(a-3)$, the LCD is $(a+3)(a-3)(a+4)$.
Multiplying by the LCD:

$$(a+3)(a-3)(a+4) \cdot \frac{2}{(a+3)(a-3)} = (a+3)(a-3)(a+4) \cdot \frac{3}{(a+4)(a-3)}$$

$$2(a+4) = 3(a+3)$$

$$2a + 8 = 3a + 9$$

$$-a + 8 = 9$$

$$-a = 1$$

$$a = -1$$

Since $a = -1$ checks in the original equation, the solution is $a = -1$.

39. Multiplying both sides of the equation by $x^2 - 4x - 5 = (x-5)(x+1)$:

$$(x-5)(x+1)\left(\frac{3x}{x-5} - \frac{2x}{x+1}\right) = (x-5)(x+1) \cdot \frac{-42}{(x-5)(x+1)}$$

$$3x(x+1) - 2x(x-5) = -42$$

$$3x^2 + 3x - 2x^2 + 10x = -42$$

$$x^2 + 13x + 42 = 0$$

$$(x+7)(x+6) = 0$$

$$x = -7, -6$$

Both $x = -7$ and $x = -6$ check in the original equation.

41. Multiplying both sides of the equation by $x^2 + 5x + 6 = (x+2)(x+3)$:

$$(x+2)(x+3)\left(\frac{2x}{x+2}\right) = (x+2)(x+3)\left(\frac{x}{x+3} - \frac{3}{x^2+5x+6}\right)$$

$$2x(x+3) = x(x+2) - 3$$

$$2x^2 + 6x = x^2 + 2x - 3$$

$$x^2 + 4x + 3 = 0$$

$$(x+3)(x+1) = 0$$

$$x = -3, -1$$

Since $x = -3$ does not check in the original equation, the solution is $x = -1$.

43. **a.** Solving the equation:

$$5x - 1 = 0$$
$$5x = 1$$
$$x = \frac{1}{5}$$

b. Solving the equation:

$$\frac{5}{x} - 1 = 0$$
$$x\left(\frac{5}{x} - 1\right) = x(0)$$
$$5 - x = 0$$
$$x = 5$$

c. Solving the equation:

$$\frac{x}{5} - 1 = \frac{2}{3}$$
$$15\left(\frac{x}{5} - 1\right) = 15\left(\frac{2}{3}\right)$$
$$3x - 15 = 10$$
$$3x = 25$$
$$x = \frac{25}{3}$$

d. Solving the equation:

$$\frac{5}{x} - 1 = \frac{2}{3}$$
$$3x\left(\frac{5}{x} - 1\right) = 3x\left(\frac{2}{3}\right)$$
$$15 - 3x = 2x$$
$$15 = 5x$$
$$x = 3$$

e. Solving the equation:

$$\frac{5}{x^2} + 5 = \frac{26}{x}$$
$$x^2\left(\frac{5}{x^2} + 5\right) = x^2\left(\frac{26}{x}\right)$$
$$5 + 5x^2 = 26x$$
$$5x^2 - 26x + 5 = 0$$
$$(5x - 1)(x - 5) = 0$$
$$x = \frac{1}{5}, 5$$

45. **a.** Dividing: $\dfrac{7}{a^2 - 5a - 6} \div \dfrac{a+2}{a+1} = \dfrac{7}{(a-6)(a+1)} \cdot \dfrac{a+1}{a+2} = \dfrac{7}{(a-6)(a+2)}$

b. Adding:

$$\frac{7}{a^2 - 5a - 6} + \frac{a+2}{a+1} = \frac{7}{(a-6)(a+1)} + \frac{a+2}{a+1} \cdot \frac{a-6}{a-6}$$
$$= \frac{7}{(a-6)(a+1)} + \frac{a^2 - 4a - 12}{(a-6)(a+1)}$$
$$= \frac{a^2 - 4a - 5}{(a-6)(a+1)}$$
$$= \frac{(a-5)(a+1)}{(a-6)(a+1)}$$
$$= \frac{a-5}{a-6}$$

c.   Solving the equation:

$$\frac{7}{a^2 - 5a - 6} + \frac{a+2}{a+1} = 2$$

$$\frac{7}{(a-6)(a+1)} + \frac{a+2}{a+1} = 2$$

$$(a-6)(a+1)\left(\frac{7}{(a-6)(a+1)} + \frac{a+2}{a+1}\right) = 2(a-6)(a+1)$$

$$7 + (a-6)(a+2) = 2(a-6)(a+1)$$

$$7 + a^2 - 4a - 12 = 2a^2 - 10a - 12$$

$$0 = a^2 - 6a - 7$$

$$0 = (a-7)(a+1)$$

$$a = -1, 7$$

Upon checking, only $a = 7$ checks in the original equation.

47.   Multiplying both sides of the equation by $2x$:

$$2x\left(\frac{1}{x} + \frac{1}{2x}\right) = 2x\left(\frac{9}{2}\right)$$

$$2 + 1 = 9x$$

$$9x = 3$$

$$x = \frac{1}{3}$$

49.   Multiplying both sides of the equation by $30x$:

$$30x\left(\frac{1}{10} - \frac{1}{15}\right) = 30x\left(\frac{1}{x}\right)$$

$$3x - 2x = 30$$

$$x = 30$$

51.   Evaluating when $x = -6$: $y = -\dfrac{6}{-6} = 1$ 53.   Evaluating when $x = 2$: $y = -\dfrac{6}{2} = -3$

6.5 Applications

1.   Let x and $3x$ represent the two numbers. The equation is:

$$\frac{1}{x} + \frac{1}{3x} = \frac{16}{3}$$

$$3x\left(\frac{1}{x} + \frac{1}{3x}\right) = 3x\left(\frac{16}{3}\right)$$

$$3 + 1 = 16x$$

$$16x = 4$$

$$x = \frac{1}{4}$$

$$3x = \frac{3}{4}$$

The numbers are $\dfrac{1}{4}$ and $\dfrac{3}{4}$.

3. Let x represent the number. The equation is:

$$x + \frac{1}{x} = \frac{13}{6}$$

$$6x\left(x + \frac{1}{x}\right) = 6x\left(\frac{13}{6}\right)$$

$$6x^2 + 6 = 13x$$

$$6x^2 - 13x + 6 = 0$$

$$(3x - 2)(2x - 3) = 0$$

$$x = \frac{2}{3}, \frac{3}{2}$$

The number is either $\frac{2}{3}$ or $\frac{3}{2}$.

5. Let x represent the number. The equation is:

$$\frac{7 + x}{9 + x} = \frac{5}{7}$$

$$7(9 + x) \cdot \frac{7 + x}{9 + x} = 7(9 + x) \cdot \frac{5}{7}$$

$$7(7 + x) = 5(9 + x)$$

$$49 + 7x = 45 + 5x$$

$$49 + 2x = 45$$

$$2x = -4$$

$$x = -2$$

The number is –2.

7. Let x and $x + 2$ represent the two integers. The equation is:

$$\frac{1}{x} + \frac{1}{x + 2} = \frac{5}{12}$$

$$12x(x + 2)\left(\frac{1}{x} + \frac{1}{x + 2}\right) = 12x(x + 2)\left(\frac{5}{12}\right)$$

$$12(x + 2) + 12x = 5x(x + 2)$$

$$12x + 24 + 12x = 5x^2 + 10x$$

$$0 = 5x^2 - 14x - 24$$

$$(5x + 6)(x - 4) = 0$$

$$x = 4 \quad \left(x = -\frac{6}{5} \text{ is impossible}\right)$$

$$x + 2 = 6$$

The integers are 4 and 6.

9. Let x represent the rate of the boat in still water:

	d	r	t
Upstream	26	$x - 3$	$\dfrac{26}{x - 3}$
Downstream	38	$x + 3$	$\dfrac{38}{x + 3}$

The equation is:

$$\frac{26}{x - 3} = \frac{38}{x + 3}$$

$$(x + 3)(x - 3) \cdot \frac{26}{x - 3} = (x + 3)(x - 3) \cdot \frac{38}{x + 3}$$

$$26(x + 3) = 38(x - 3)$$

$$26x + 78 = 38x - 114$$

$$-12x + 78 = -114$$

$$-12x = -192$$

$$x = 16$$

The speed of the boat in still water is 16 mph.

11. Let x represent the plane speed in still air:

	d	r	t
Against Wind	140	$x-20$	$\dfrac{140}{x-20}$
With Wind	160	$x+20$	$\dfrac{160}{x+20}$

The equation is:

$$\frac{140}{x-20}=\frac{160}{x+20}$$
$$(x+20)(x-20)\bullet\frac{140}{x-20}=(x+20)(x-20)\bullet\frac{160}{x+20}$$
$$140(x+20)=160(x-20)$$
$$140x+2800=160x-3200$$
$$-20x+2800=-3200$$
$$-20x=-6000$$
$$x=300$$

The plane speed in still air is 300 mph.

13. Let x and $x+20$ represent the rates of each plane:

	d	r	t
Plane 1	285	$x+20$	$\dfrac{285}{x+20}$
Plane 2	255	x	$\dfrac{255}{x}$

The equation is:

$$\frac{285}{x+20}=\frac{255}{x}$$
$$x(x+20)\bullet\frac{285}{x+20}=x(x+20)\bullet\frac{255}{x}$$
$$285x=255(x+20)$$
$$285x=255x+5100$$
$$30x=5100$$
$$x=170$$
$$x+20=190$$

The plane speeds are 170 mph and 190 mph.

15. Let x represent her rate downhill:

	d	r	t
Level Ground	2	$x-3$	$\dfrac{2}{x-3}$
Downhill	6	x	$\dfrac{6}{x}$

The equation is:
$$\frac{2}{x-3}+\frac{6}{x}=1$$
$$x(x-3)\left(\frac{2}{x-3}+\frac{6}{x}\right)=x(x-3)\bullet 1$$
$$2x+6(x-3)=x(x-3)$$
$$2x+6x-18=x^2-3x$$
$$8x-18=x^2-3x$$
$$0=x^2-11x+18$$
$$0=(x-2)(x-9)$$
$$x=9\quad (x=2 \text{ is impossible})$$

Tina runs 9 mph on the downhill part of the course.

17. Let x represent her rate on level ground:

	d	r	t
Level Ground	4	x	$\dfrac{4}{x}$
Downhill	5	$x+2$	$\dfrac{5}{x+2}$

The equation is:
$$\frac{4}{x}+\frac{5}{x+2}=1$$
$$x(x+2)\left(\frac{4}{x}+\frac{5}{x+2}\right)=x(x+2)\bullet 1$$
$$4(x+2)+5x=x(x+2)$$
$$4x+8+5x=x^2+2x$$
$$9x+8=x^2+2x$$
$$0=x^2-7x-8$$
$$0=(x-8)(x+1)$$
$$x=8\quad (x=-1 \text{ is impossible})$$

Jerri jogs 8 mph on level ground.

19. Let t represent the time to fill the pool with both pipes left open. The equation is:
$$\frac{1}{12}-\frac{1}{15}=\frac{1}{t}$$
$$60t\left(\frac{1}{12}-\frac{1}{15}\right)=60t\bullet\frac{1}{t}$$
$$5t-4t=60$$
$$t=60$$

It will take 60 hours to fill the pool with both pipes left open.

21.   Let t represent the time to fill the bathtub with both faucets open. The equation is:

$$\frac{1}{10}+\frac{1}{12}=\frac{1}{t}$$

$$60t\left(\frac{1}{10}+\frac{1}{12}\right)=60t\cdot\frac{1}{t}$$

$$6t+5t=60$$

$$11t=60$$

$$t=\frac{60}{11}=5\frac{5}{11}$$

It will take $5\frac{5}{11}$ minutes to fill the tub with both faucets open.

23.   Let t represent the time to fill the sink with both the faucet and the drain left open. The equation is:

$$\frac{1}{3}-\frac{1}{4}=\frac{1}{t}$$

$$12t\left(\frac{1}{3}-\frac{1}{4}\right)=12t\cdot\frac{1}{t}$$

$$4t-3t=12$$

$$t=12$$

It will take 12 minutes for the sink to overflow with both the faucet and drain left open.

25.   Sketching the graph: **27.**   Sketching the graph:

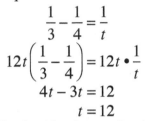

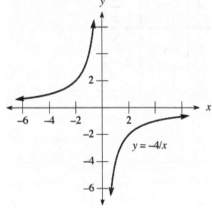

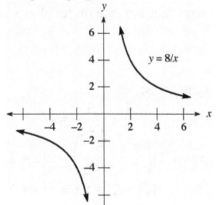

29.   Sketching the graph:

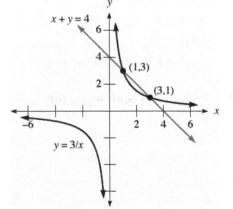

The intersection points are (1,3) and (3,1).

31. Simplifying: $\dfrac{1}{2} \div \dfrac{2}{3} = \dfrac{1}{2} \cdot \dfrac{3}{2} = \dfrac{3}{4}$

33. Simplifying: $1 + \dfrac{1}{2} = \dfrac{2}{2} + \dfrac{1}{2} = \dfrac{3}{2}$

35. Simplifying: $y^5 \cdot \dfrac{2x^3}{y^2} = \dfrac{2x^3 y^5}{y^2} = 2x^3 y^3$

37. Simplifying: $\dfrac{2x^3}{y^2} \cdot \dfrac{y^5}{4x} = \dfrac{2x^3 y^5}{4xy^2} = \dfrac{x^2 y^3}{2}$

39. Factoring: $x^2 y + x = x(xy + 1)$

41. Reducing the fraction: $\dfrac{2x^3 y^2}{4x} = \dfrac{x^2 y^2}{2}$

43. Reducing the fraction: $\dfrac{x^2 - 4}{x^2 - x - 6} = \dfrac{(x+2)(x-2)}{(x+2)(x-3)} = \dfrac{x-2}{x-3}$

6.6 Complex Fractions

1. Simplifying the complex fraction: $\dfrac{\frac{3}{4}}{\frac{7}{8}} = \dfrac{\frac{3}{4} \cdot 8}{\frac{7}{8} \cdot 8} = \dfrac{6}{1} = 6$

3. Simplifying the complex fraction: $\dfrac{\frac{2}{3}}{4} = \dfrac{\frac{2}{3} \cdot 3}{4 \cdot 3} = \dfrac{2}{12} = \dfrac{1}{6}$

5. Simplifying the complex fraction: $\dfrac{\frac{x^2}{y}}{\frac{x}{y^3}} = \dfrac{\frac{x^2}{y} \cdot y^3}{\frac{x}{y^3} \cdot y^3} = \dfrac{x^2 y^2}{x} = xy^2$

7. Simplifying the complex fraction: $\dfrac{\frac{4x^3}{y^6}}{\frac{8x^2}{y^7}} = \dfrac{\frac{4x^3}{y^6} \cdot y^7}{\frac{8x^2}{y^7} \cdot y^7} = \dfrac{4x^3 y}{8x^2} = \dfrac{xy}{2}$

9. Simplifying the complex fraction: $\dfrac{y + \frac{1}{x}}{x + \frac{1}{y}} = \dfrac{\left(y + \frac{1}{x}\right) \cdot xy}{\left(x + \frac{1}{y}\right) \cdot xy} = \dfrac{xy^2 + y}{x^2 y + x} = \dfrac{y(xy + 1)}{x(xy + 1)} = \dfrac{y}{x}$

11. Simplifying the complex fraction: $\dfrac{1 + \frac{1}{a}}{1 - \frac{1}{a}} = \dfrac{\left(1 + \frac{1}{a}\right) \cdot a}{\left(1 - \frac{1}{a}\right) \cdot a} = \dfrac{a+1}{a-1}$

13. Simplifying the complex fraction: $\dfrac{\frac{x+1}{x^2-9}}{\frac{2}{x+3}} = \dfrac{\frac{x+1}{(x+3)(x-3)} \cdot (x+3)(x-3)}{\frac{2}{x+3} \cdot (x+3)(x-3)} = \dfrac{x+1}{2(x-3)}$

15. Simplifying the complex fraction: $\dfrac{\frac{1}{a+2}}{\frac{1}{a^2-a-6}} = \dfrac{\frac{1}{a+2} \cdot (a-3)(a+2)}{\frac{1}{(a-3)(a+2)} \cdot (a-3)(a+2)} = \dfrac{a-3}{1} = a-3$

17. Simplifying the complex fraction:

$$\frac{1-\dfrac{9}{y^2}}{1-\dfrac{1}{y}-\dfrac{6}{y^2}}=\frac{\left(1-\dfrac{9}{y^2}\right)\bullet y^2}{\left(1-\dfrac{1}{y}-\dfrac{6}{y^2}\right)\bullet y^2}=\frac{y^2-9}{y^2-y-6}=\frac{(y+3)(y-3)}{(y+2)(y-3)}=\frac{y+3}{y+2}$$

19. Simplifying the complex fraction: $\dfrac{\dfrac{1}{y}+\dfrac{1}{x}}{\dfrac{1}{xy}}=\dfrac{\left(\dfrac{1}{y}+\dfrac{1}{x}\right)\bullet xy}{\left(\dfrac{1}{xy}\right)\bullet xy}=\dfrac{x+y}{1}=x+y$

21. Simplifying the complex fraction: $\dfrac{1-\dfrac{1}{a^2}}{1-\dfrac{1}{a}}=\dfrac{\left(1-\dfrac{1}{a^2}\right)\bullet a^2}{\left(1-\dfrac{1}{a}\right)\bullet a^2}=\dfrac{a^2-1}{a^2-a}=\dfrac{(a+1)(a-1)}{a(a-1)}=\dfrac{a+1}{a}$

23. Simplifying the complex fraction: $\dfrac{\dfrac{1}{10x}-\dfrac{y}{10x^2}}{\dfrac{1}{10}-\dfrac{y}{10x}}=\dfrac{\left(\dfrac{1}{10x}-\dfrac{y}{10x^2}\right)\bullet 10x^2}{\left(\dfrac{1}{10}-\dfrac{y}{10x}\right)\bullet 10x^2}=\dfrac{x-y}{x^2-xy}=\dfrac{1(x-y)}{x(x-y)}=\dfrac{1}{x}$

25. Simplifying the complex fraction:

$$\frac{\dfrac{1}{a+1}+2}{\dfrac{1}{a+1}+3}=\frac{\left(\dfrac{1}{a+1}+2\right)\bullet(a+1)}{\left(\dfrac{1}{a+1}+3\right)\bullet(a+1)}=\frac{1+2(a+1)}{1+3(a+1)}=\frac{1+2a+2}{1+3a+3}=\frac{2a+3}{3a+4}$$

27. Simplifying each parenthesis first:

$$1-\frac{1}{x}=\frac{x}{x}-\frac{1}{x}=\frac{x-1}{x}$$

$$1-\frac{1}{x+1}=\frac{x+1}{x+1}-\frac{1}{x+1}=\frac{x}{x+1}$$

$$1-\frac{1}{x+2}=\frac{x+2}{x+2}-\frac{1}{x+2}=\frac{x+1}{x+2}$$

Now performing the multiplication: $\left(1-\dfrac{1}{x}\right)\left(1-\dfrac{1}{x+1}\right)\left(1-\dfrac{1}{x+2}\right)=\dfrac{x-1}{x}\bullet\dfrac{x}{x+1}\bullet\dfrac{x+1}{x+2}=\dfrac{x-1}{x+2}$

29. Simplifying each parenthesis first:

$$1+\frac{1}{x+3}=\frac{x+3}{x+3}+\frac{1}{x+3}=\frac{x+4}{x+3}$$

$$1+\frac{1}{x+2}=\frac{x+2}{x+2}+\frac{1}{x+2}=\frac{x+3}{x+2}$$

$$1+\frac{1}{x+1}=\frac{x+1}{x+1}+\frac{1}{x+1}=\frac{x+2}{x+1}$$

Now performing the multiplication:

$$\left(1+\frac{1}{x+3}\right)\left(1+\frac{1}{x+2}\right)\left(1+\frac{1}{x+1}\right)=\frac{x+4}{x+3}\bullet\frac{x+3}{x+2}\bullet\frac{x+2}{x+1}=\frac{x+4}{x+1}$$

31. Simplifying each term in the sequence:

$$2 + \frac{1}{2+1} = 2 + \frac{1}{3} = \frac{6}{3} + \frac{1}{3} = \frac{7}{3}$$

$$2 + \cfrac{1}{2 + \cfrac{1}{2+1}} = 2 + \frac{1}{\frac{7}{3}} = 2 + \frac{3}{7} = \frac{14}{7} + \frac{3}{7} = \frac{17}{7}$$

$$2 + \cfrac{2}{2 + \cfrac{1}{2 + \cfrac{1}{2+1}}} = 2 + \frac{1}{\frac{17}{7}} = 2 + \frac{7}{17} = \frac{34}{17} + \frac{7}{17} = \frac{41}{17}$$

33. Completing the table:

Number x	Reciprocal $\frac{1}{x}$	Quotient $\dfrac{x}{1/x}$	Square x^2
1	1	1	1
2	$\frac{1}{2}$	4	4
3	$\frac{1}{3}$	9	9
4	$\frac{1}{4}$	16	16

35. Completing the table:

Number x	Reciprocal $\frac{1}{x}$	Sum $1+\frac{1}{x}$	Quotient $\dfrac{1+\frac{1}{x}}{\frac{1}{x}}$
1	1	2	2
2	$\frac{1}{2}$	$\frac{3}{2}$	3
3	$\frac{1}{3}$	$\frac{4}{3}$	4
4	$\frac{1}{4}$	$\frac{5}{4}$	5

37. Solving the equation:

$$21 = 6x$$
$$x = \frac{21}{6} = \frac{7}{2}$$

39. Solving the equation:

$$x^2 + x = 6$$
$$x^2 + x - 6 = 0$$
$$(x+3)(x-2) = 0$$
$$x = -3, 2$$

6.7 Proportions

1. Solving the proportion:
$$\frac{x}{2} = \frac{6}{12}$$
$$12x = 12$$
$$x = 1$$

3. Solving the proportion:
$$\frac{2}{5} = \frac{4}{x}$$
$$2x = 20$$
$$x = 10$$

5. Solving the proportion:
$$\frac{10}{20} = \frac{20}{x}$$
$$10x = 400$$
$$x = 40$$

7. Solving the proportion:
$$\frac{a}{3} = \frac{5}{12}$$
$$12a = 15$$
$$a = \frac{15}{12} = \frac{5}{4}$$

9. Solving the proportion:
$$\frac{2}{x} = \frac{6}{7}$$
$$6x = 14$$
$$x = \frac{14}{6} = \frac{7}{3}$$

11. Solving the proportion:
$$\frac{x+1}{3} = \frac{4}{x}$$
$$x^2 + x = 12$$
$$x^2 + x - 12 = 0$$
$$(x+4)(x-3) = 0$$
$$x = -4, 3$$

13. Solving the proportion:
$$\frac{x}{2} = \frac{8}{x}$$
$$x^2 = 16$$
$$x^2 - 16 = 0$$
$$(x+4)(x-4) = 0$$
$$x = -4, 4$$

15. Solving the proportion:
$$\frac{4}{a+2} = \frac{a}{2}$$
$$a^2 + 2a = 8$$
$$a^2 + 2a - 8 = 0$$
$$(a+4)(a-2) = 0$$
$$a = -4, 2$$

17. Solving the proportion:
$$\frac{1}{x} = \frac{x-5}{6}$$
$$x^2 - 5x = 6$$
$$x^2 - 5x - 6 = 0$$
$$(x-6)(x+1) = 0$$
$$x = -1, 6$$

19. Comparing hits to games, the proportion is:
$$\frac{6}{18} = \frac{x}{45}$$
$$18x = 270$$
$$x = 15$$
He will get 15 hits in 45 games.

21. Comparing ml alcohol to ml water, the proportion is:

$$\frac{12}{16} = \frac{x}{28}$$
$$16x = 336$$
$$x = 21$$

The solution will have 21 ml of alcohol.

23. Comparing grams of fat to total grams, the proportion is:

$$\frac{13}{100} = \frac{x}{350}$$
$$100x = 4550$$
$$x = 45.5$$

There are 45.5 grams of fat in 350 grams of ice cream.

25. Comparing inches on the map to actual miles, the proportion is:

$$\frac{3.5}{100} = \frac{x}{420}$$
$$100x = 1470$$
$$x = 14.7$$

They are 14.7 inches apart on the map.

27. Comparing miles to hours, the proportion is:

$$\frac{245}{5} = \frac{x}{7}$$
$$5x = 1715$$
$$x = 343$$

He will travel 343 miles.

29. Evaluating when $x = 4$: $y = 5(4) = 20$

31. Evaluating when $x = 10$: $y = \frac{20}{10} = 2$

33. Substituting $y = 50$:

$$50 = 2x^2$$
$$2x^2 - 50 = 0$$
$$2(x^2 - 25) = 0$$
$$2(x + 5)(x - 5) = 0$$
$$x = -5, 5$$

35. Substituting $y = 15$ and $x = 3$:

$$15 = K \bullet 3$$
$$K = 5$$

37. Substituting $y = 32$ and $x = 4$:

$$32 = K(4)^2$$
$$32 = 16K$$
$$K = 2$$

6.8 Variation

1.   The variation equation is $y = Kx$. Finding K:
$$10 = K \bullet 5$$
$$K = 2$$
So $y = 2x$. Substituting $x = 4$: $y = 2 \bullet 4 = 8$

3.   The variation equation is $y = Kx$. Finding K:
$$39 = K \bullet 3$$
$$K = 13$$
So $y = 13x$. Substituting $x = 10$: $y = 13 \bullet 10 = 130$

5.   The variation equation is $y = Kx$. Finding K:
$$-24 = K \bullet 4$$
$$K = -6$$
So $y = -6x$. Substituting $y = -30$:
$$-6x = -30$$
$$x = 5$$

7.   The variation equation is $y = Kx$. Finding K:
$$-7 = K \bullet (-1)$$
$$K = 7$$
So $y = 7x$. Substituting $y = -21$:
$$7x = -21$$
$$x = -3$$

9.   The variation equation is $y = Kx^2$. Finding K:
$$75 = K \bullet 5^2$$
$$75 = 25K$$
$$K = 3$$
So $y = 3x^2$. Substituting $x = 1$: $y = 3 \bullet 1^2 = 3 \bullet 1 = 3$

11.   The variation equation is $y = Kx^2$. Finding K:
$$48 = K \bullet 4^2$$
$$48 = 16K$$
$$K = 3$$
So $y = 3x^2$. Substituting $x = 9$: $y = 3 \bullet 9^2 = 3 \bullet 81 = 243$

13.   The variation equation is $y = \dfrac{K}{x}$. Finding K:
$$5 = \dfrac{K}{2}$$
$$K = 10$$
So $y = \dfrac{10}{x}$. Substituting $x = 5$: $y = \dfrac{10}{5} = 2$

15.   The variation equation is $y = \dfrac{K}{x}$. Finding K:
$$2 = \dfrac{K}{1}$$
$$K = 2$$
So $y = \dfrac{2}{x}$. Substituting $x = 4$: $y = \dfrac{2}{4} = \dfrac{1}{2}$

17. The variation equation is $y = \dfrac{K}{x}$. Finding K:

$$5 = \frac{K}{3}$$
$$K = 15$$

So $y = \dfrac{15}{x}$. Substituting $y = 15$:

$$\frac{15}{x} = 15$$
$$15 = 15x$$
$$x = 1$$

19. The variation equation is $y = \dfrac{K}{x}$. Finding K:

$$10 = \frac{K}{10}$$
$$K = 100$$

So $y = \dfrac{100}{x}$. Substituting $y = 100$:

$$\frac{100}{x} = 20$$
$$100 = 20x$$
$$x = 5$$

21. The variation equation is $y = \dfrac{K}{x^2}$. Finding K:

$$4 = \frac{K}{5^2}$$
$$4 = \frac{K}{25}$$
$$K = 100$$

So $y = \dfrac{100}{x^2}$. Substituting $x = 2$: $y = \dfrac{100}{2^2} = \dfrac{100}{4} = 25$

23. The variation equation is $y = \dfrac{K}{x^2}$. Finding K:

$$4 = \frac{K}{3^2}$$
$$4 = \frac{K}{9}$$
$$K = 36$$

So $y = \dfrac{36}{x^2}$. Substituting $x = 2$: $y = \dfrac{36}{2^2} = \dfrac{36}{4} = 9$

25. The variation equation is $t = Kd$. Finding K:

$$42 = K \cdot 2$$
$$K = 21$$

So $t = 21d$. Substituting $d = 4$: $t = 21 \cdot 4 = 84$ pounds

27. The variation equation is $P = KI^2$. Finding K:

$$30 = K \cdot 2^2$$
$$30 = 4K$$
$$K = \frac{15}{2}$$

So $P = \dfrac{15}{2}I^2$. Substituting $I = 7$: $P = \dfrac{15}{2} \cdot 7^2 = \dfrac{15}{2} \cdot 49 = 367.5$

29. The variation equation is $M = Kh$. Finding K:

$$157 = K \cdot 20$$
$$K = 7.85$$

So $M = 7.85h$. Substituting $h = 30$: $M = 7.85 \cdot 30 = \$235.50$

31.   The variation equation is $F = \dfrac{K}{d^2}$. Finding K:

$$150 = \frac{K}{4000^2}$$
$$150 = \frac{K}{1.6 \times 10^7}$$
$$K = 2.4 \times 10^9$$

So $F = \dfrac{2.4 \times 10^9}{d^2}$. Substituting $d = 5000$: $F = \dfrac{2.4 \times 10^9}{(5000)^2} = \dfrac{2.4 \times 10^9}{2.5 \times 10^7} = 96$ pounds

33.   The variation equation is $I = \dfrac{K}{R}$. Finding K:

$$30 = \frac{K}{2}$$
$$K = 60$$

So $I = \dfrac{60}{R}$. Substituting $R = 5$: $I = \dfrac{60}{5} = 12$ amps

35.   Substituting $x = -2$, $x = 0$, and $x = 2$:

$$y = \frac{1}{2}(-2) + 3 = -1 + 3 = 2 \qquad y = \frac{1}{2}(0) + 3 = 0 + 3 = 3 \qquad y = \frac{1}{2}(2) + 3 = 1 + 3 = 4$$

The ordered pairs are $(-2,2)$, $(0,3)$, and $(2,4)$.

37.   Graphing the line:

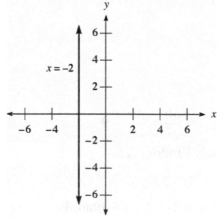

39.   Computing the slope: $m = \dfrac{1 - 5}{0 - 2} = \dfrac{-4}{-2} = 2$

41.   Using the point-slope formula:

$$y - 1 = \frac{1}{2}\left(x - (-2)\right)$$
$$y - 1 = \frac{1}{2}(x + 2)$$
$$y - 1 = \frac{1}{2}x + 1$$
$$y = \frac{1}{2}x + 2$$

43. Computing the slope: $m = \dfrac{1-5}{0-2} = \dfrac{-4}{-2} = 2$. Using the point-slope formula:

$$y - 1 = 2(x - 0)$$
$$y - 1 = 2x$$
$$y = 2x + 1$$

Chapter 6 Test

See www.mathtv.com for video solutions to all problems in this chapter test.

162

Chapter 7
Systems of Linear Equations

7.1 Solving Linear Systems by Graphing

1. Graphing both lines:

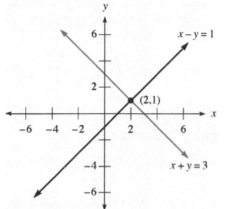

The intersection point is $(2,1)$.

3. Graphing both lines:

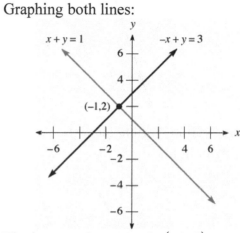

The intersection point is $(-1,2)$.

5. Graphing both lines:

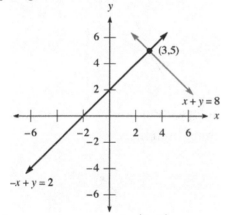

The intersection point is $(3,5)$.

7. Graphing both lines:

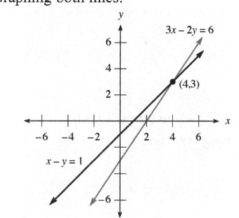

The intersection point is $(4,3)$.

9.   Graphing both lines:

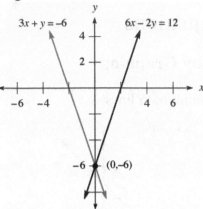

The intersection point is $(0,-6)$.

11. Graphing both lines:

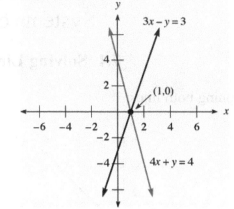

The intersection point is $(1,0)$.

13. Graphing both lines:

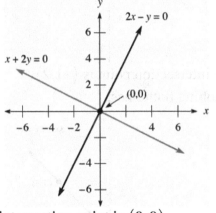

The intersection point is $(0,0)$.

15. Graphing both lines:

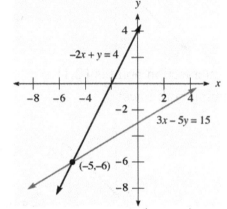

The intersection point is $(-5,-6)$.

17. Graphing both lines:

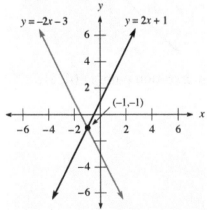

The intersection point is $(-1,-1)$.

19. Graphing both lines:

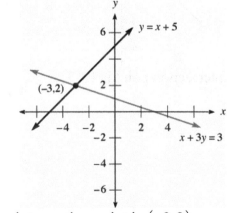

The intersection point is $(-3,2)$.

21. Graphing both lines:

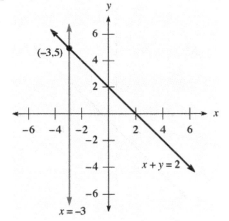

The intersection point is $(-3,5)$.

23. Graphing both lines:

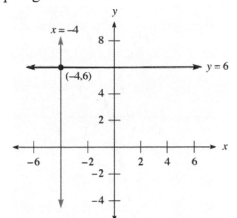

The intersection point is $(-4,6)$.

25. Graphing both lines:

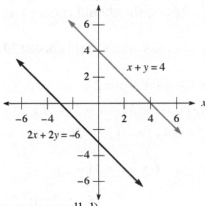

There is no intersection (the lines are parallel).

27. Graphing both lines:

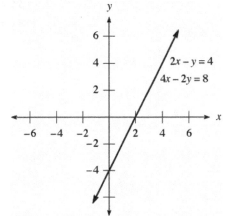

The system is dependent (both lines are the same, they coincide).

29.   Graphing both lines:

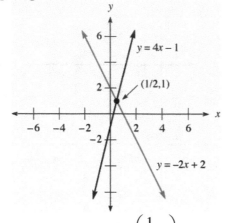

The intersection point is $\left(\dfrac{1}{2},1\right)$.

31.   Graphing both lines:

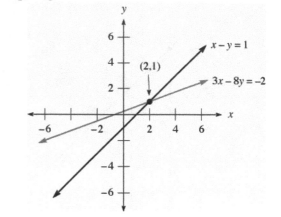

The intersection point is $(2,1)$.

33.   **a.**   If Jane worked 25 hours, she would earn the same amount at each position.
     b.   If Jane worked less than 20 hours, she should choose Gigi's since she earns more in that position.
     c.   If Jane worked more than 30 hours, she should choose Marcy's since she earns more in that position.

35.   Simplifying: $(x+y)+(x-y)=x+y+x-y=2x$

37.   Simplifying: $3(2x-y)+(x+3y)=6x-3y+x+3y=7x$

39.   Simplifying: $-4(3x+5y)+5(5x+4y)=-12x-20y+25x+20y=13x$

41.   Simplifying: $6\left(\dfrac{1}{2}x-\dfrac{1}{3}y\right)=6\bullet\dfrac{1}{2}x-6\bullet\dfrac{1}{3}y=3x-2y$

43.   Substituting $x=3$:

$$3+y=4$$
$$y=1$$

45.   Substituting $x=3$:

$$3+3y=3$$
$$3y=0$$
$$y=0$$

47.   Substituting $x=6$:

$$3(6)+5y=-7$$
$$18+5y=-7$$
$$5y=-25$$
$$y=-5$$

7.2 The Elimination Method

1. Adding the two equations:
$$2x = 4$$
$$x = 2$$
Substituting into the first equation:
$$2 + y = 3$$
$$y = 1$$
The solution is $(2, 1)$.

3. Adding the two equations:
$$2y = 14$$
$$y = 7$$
Substituting into the first equation:
$$x + 7 = 10$$
$$x = 3$$
The solution is $(3, 7)$.

5. Adding the two equations:
$$-2y = 10$$
$$y = -5$$
Substituting into the first equation:
$$x - (-5) = 7$$
$$x + 5 = 7$$
$$x = 2$$
The solution is $(2, -5)$.

7. Adding the two equations:
$$4x = -4$$
$$x = -1$$
Substituting into the first equation:
$$-1 + y = -1$$
$$y = 0$$
The solution is $(-1, 0)$.

9. Adding the two equations:
$$0 = 0$$
The lines coincide (the system is dependent).

11. Multiplying the first equation by 2:
$$6x - 2y = 8$$
$$2x + 2y = 24$$
Adding the two equations:
$$8x = 32$$
$$x = 4$$
Substituting into the first equation:
$$3(4) - y = 4$$
$$12 - y = 4$$
$$-y = -8$$
$$y = 8$$
The solution is $(4, 8)$.

13. Multiplying the second equation by -3:
$$5x - 3y = -2$$
$$-30x + 3y = -3$$
Adding the two equations:
$$-25x = -5$$
$$x = \frac{1}{5}$$
Substituting into the first equation:
$$5\left(\frac{1}{5}\right) - 3y = -2$$
$$1 - 3y = -2$$
$$-3y = -3$$
$$y = 1$$
The solution is $\left(\frac{1}{5}, 1\right)$.

15. Multiplying the second equation by 4:
$$11x - 4y = 11$$
$$20x + 4y = 20$$
Adding the two equations:
$$31x = 31$$
$$x = 1$$

17. Multiplying the second equation by 3:
$$3x - 5y = 7$$
$$-3x + 3y = -3$$
Adding the two equations:
$$-2y = 4$$
$$y = -2$$

Substituting into the second equation:
$$5(1) + y = 5$$
$$5 + y = 5$$
$$y = 0$$
The solution is $(1,0)$.

19. Multiplying the first equation by –2:
$$2x + 16y = 2$$
$$-2x + 4y = 13$$
Adding the two equations:
$$20y = 15$$
$$y = \frac{3}{4}$$
Substituting into the first equation:
$$-x - 8\left(\frac{3}{4}\right) = -1$$
$$-x - 6 = -1$$
$$-x = 5$$
$$x = -5$$
The solution is $\left(-5, \frac{3}{4}\right)$.

23. Adding the two equations:
$$8x = -24$$
$$x = -3$$
Substituting into the second equation:
$$2(-3) + y = -16$$
$$-6 + y = -16$$
$$y = -10$$
The solution is $(-3, -10)$.

25. Multiplying the second equation by 3:
$$x + 3y = 9$$
$$6x - 3y = 12$$
Adding the two equations:
$$7x = 21$$
$$x = 3$$
Substituting into the first equation:
$$3 + 3y = 9$$
$$3y = 6$$
$$y = 2$$

The solution is $(3,2)$.

Substituting into the second equation:
$$-x - 2 = -1$$
$$-x = 1$$
$$x = -1$$
The solution is $(-1, -2)$.

21. Multiplying the first equation by 2:
$$-6x - 2y = 14$$
$$6x + 7y = 11$$
Adding the two equations:
$$5y = 25$$
$$y = 5$$
Substituting into the first equation:
$$-3x - 5 = 7$$
$$-3x = 12$$
$$x = -4$$
The solution is $(-4, 5)$.

27. Multiplying the second equation by 2:
$$x - 6y = 3$$
$$8x + 6y = 42$$
Adding the two equations:
$$9x = 45$$
$$x = 5$$
Substituting into the second equation:
$$4(5) + 3y = 21$$
$$20 + 3y = 21$$
$$3y = 1$$
$$y = \frac{1}{3}$$
The solution is $\left(5, \frac{1}{3}\right)$.

29. Multiplying the second equation by -3:
$$2x+9y=2$$
$$-15x-9y=24$$
Adding the two equations:
$$-13x=26$$
$$x=-2$$
Substituting into the first equation:
$$2(-2)+9y=2$$
$$-4+9y=2$$
$$9y=6$$
$$y=\frac{2}{3}$$
The solution is $\left(-2,\frac{2}{3}\right)$.

31. To clear each equation of fractions, multiply the first equation by 12 and the second equation by 6:
$$12\left(\frac{1}{3}x+\frac{1}{4}y\right)=12\left(\frac{7}{6}\right) \qquad\qquad 6\left(\frac{3}{2}x-\frac{1}{3}y\right)=6\left(\frac{7}{3}\right)$$
$$4x+3y=14 \qquad\qquad\qquad 9x-2y=14$$
The system of equations is:
$$4x+3y=14$$
$$9x-2y=14$$
Multiplying the first equation by 2 and the second equation by 3:
$$8x+6y=28$$
$$27x-6y=42$$
Adding the two equations:
$$35x=70$$
$$x=2$$
Substituting into $4x+3y=14$:
$$4(2)+3y=14$$
$$8+3y=14$$
$$3y=6$$
$$y=2$$
The solution is $(2,2)$.

33. Multiplying the first equation by -2:
$$-6x-4y=2$$
$$6x+4y=0$$
Adding the two equations:
$$0=2$$
Since this statement is false, the two lines are parallel, so the system has no solution.

35.   Multiplying the first equation by 2 and the second equation by 3:
$$22x + 12y = 34$$
$$15x - 12y = 3$$
Adding the two equations:
$$37x = 37$$
$$x = 1$$
Substituting into the second equation:
$$5(1) - 4y = 1$$
$$5 - 4y = 1$$
$$-4y = -4$$
$$y = 1$$
The solution is $(1,1)$.

37.   To clear each equation of fractions, multiply the first equation by 6 and the second equation by 6:
$$6\left(\frac{1}{2}x + \frac{1}{6}y\right) = 6\left(\frac{1}{3}\right) \qquad\qquad 6\left(-x - \frac{1}{3}y\right) = 6\left(-\frac{1}{6}\right)$$
$$3x + y = 2 \qquad\qquad\qquad -6x - 2y = -1$$
The system of equations is:
$$3x + y = 2$$
$$-6x - 2y = -1$$
Multiplying the first equation by 2:
$$6x + 2y = 4$$
$$-6x - 2y = -1$$
Adding the two equations:
$$0 = 3$$
Since this statement is false, the two lines are parallel, so the system has no solution.

39.   Multiplying the second equation by 100 (to eliminate decimals):
$$x + y = 22$$
$$5x + 10y = 170$$
Multiplying the first equation by –5:
$$-5x - 5y = -110$$
$$5x + 10y = 170$$
Adding the two equations:
$$5y = 60$$
$$y = 12$$
Substituting into the first equation:
$$x + 12 = 22$$
$$x = 10$$
The solution is $(10,12)$.

41.   Solving the equation:
$$x + (2x - 1) = 2$$
$$x + 2x - 1 = 2$$
$$3x - 1 = 2$$
$$3x = 3$$
$$x = 1$$

43.   Solving the equation:
$$2(3y - 1) - 3y = 4$$
$$6y - 2 - 3y = 4$$
$$3y - 2 = 4$$
$$3y = 6$$
$$y = 2$$

45. Solving the equation:
$$4x + 2(-2x + 4) = 8$$
$$4x - 4x + 8 = 8$$
$$8 = 8$$
Since this statement is true, the solution is all real numbers.

47. Solving for x:
$$x - 3y = -1$$
$$x = 3y - 1$$

49. Solving for y: $y = 2(1) - 1 = 2 - 1 = 1$ **51.** Solving for x: $x = 3(2) - 1 = 6 - 1 = 5$

53. Substituting $x = 13$: $y = 1.5(13) + 15 = 19.5 + 15 = 34.5$

55. Substituting $x = 12$: $y = 0.75(12) + 24.95 = 9 + 24.95 = 33.95$

7.3 The Substitution Method

1. Substituting into the first equation:
$$x + (2x - 1) = 11$$
$$3x - 1 = 11$$
$$3x = 12$$
$$x = 4$$
$$y = 2(4) - 1 = 7$$
The solution is $(4, 7)$.

3. Substituting into the first equation:
$$x + (5x + 2) = 20$$
$$6x + 2 = 20$$
$$6x = 18$$
$$x = 3$$
$$y = 5(3) + 2 = 17$$
The solution is $(3, 17)$.

5. Substituting into the first equation:
$$-2x + (-4x + 8) = -1$$
$$-6x + 8 = -1$$
$$-6x = -9$$
$$x = \frac{3}{2}$$
$$y = -4\left(\frac{3}{2}\right) + 8 = -6 + 8 = 2$$
The solution is $\left(\frac{3}{2}, 2\right)$.

7. Substituting into the first equation:
$$3(-y + 6) - 2y = -2$$
$$-3y + 18 - 2y = -2$$
$$-5y + 18 = -2$$
$$-5y = -20$$
$$y = 4$$
$$x = -4 + 6 = 2$$
The solution is $(2, 4)$.

9. Substituting into the first equation:
$$5x - 4(4) = -16$$
$$5x - 16 = -16$$
$$5x = 0$$
$$x = 0$$
The solution is $(0, 4)$.

11. Substituting into the first equation:
$$5x + 4(-3x) = 7$$
$$5x - 12x = 7$$
$$-7x = 7$$
$$x = -1$$
$$y = -3(-1) = 3$$
The solution is $(-1, 3)$.

13. Solving the second equation for x:
$$x - 2y = -1$$
$$x = 2y - 1$$
Substituting into the first equation:
$$(2y - 1) + 3y = 4$$
$$5y - 1 = 4$$
$$5y = 5$$
$$y = 1$$
$$x = 2(1) - 1 = 1$$

The solution is $(1, 1)$.

15. Solving the second equation for x:
$$x - 5y = 17$$
$$x = 5y + 17$$
Substituting into the first equation:
$$2(5y + 17) + y = 1$$
$$10y + 34 + y = 1$$
$$11y + 34 = 1$$
$$11y = -33$$
$$y = -3$$
$$x = 5(-3) + 17 = 2$$

The solution is $(2, -3)$.

17. Solving the second equation for x:
$$x - 5y = -5$$
$$x = 5y - 5$$
Substituting into the first equation:
$$3(5y - 5) + 5y = -3$$
$$15y - 15 + 5y = -3$$
$$20y - 15 = -3$$
$$20y = 12$$
$$y = \frac{3}{5}$$
$$x = 5\left(\frac{3}{5}\right) - 5 = 3 - 5 = -2$$

The solution is $\left(-2, \frac{3}{5}\right)$.

19. Solving the second equation for x:
$$x - 3y = -18$$
$$x = 3y - 18$$
Substituting into the first equation:
$$5(3y - 18) + 3y = 0$$
$$15y - 90 + 3y = 0$$
$$18y - 90 = 0$$
$$18y = 90$$
$$y = 5$$
$$x = 3(5) - 18 = -3$$

The solution is $(-3, 5)$.

21. Solving the second equation for x:
$$x + 3y = 12$$
$$x = -3y + 12$$
Substituting into the first equation:
$$-3(-3y + 12) - 9y = 7$$
$$9y - 36 - 9y = 7$$
$$-36 = 7$$
Since this statement is false, there is no solution to the system. The two lines are parallel.

23. Substituting into the first equation:
$$5x - 8(2x - 5) = 7$$
$$5x - 16x + 40 = 7$$
$$-11x + 40 = 7$$
$$-11x = -33$$
$$x = 3$$
$$y = 2(3) - 5 = 1$$

The solution is $(3, 1)$.

25. Substituting into the first equation:
$$7(2y - 1) - 6y = -1$$
$$14y - 7 - 6y = -1$$
$$8y - 7 = -1$$
$$8y = 6$$
$$y = \frac{3}{4}$$
$$x = 2\left(\frac{3}{4}\right) - 1 = \frac{3}{2} - 1 = \frac{1}{2}$$

The solution is $\left(\frac{1}{2}, \frac{3}{4}\right)$.

27. Substituting into the first equation:
$$-3x + 2(3x) = 6$$
$$-3x + 6x = 6$$
$$3x = 6$$
$$x = 2$$
$$y = 3(2) = 6$$
The solution is $(2,6)$.

29. Substituting into the first equation:
$$5(y) - 6y = -4$$
$$-y = -4$$
$$y = 4$$
$$x = 4$$
The solution is $(4,4)$.

31. Substituting into the first equation:
$$3x + 3(2x - 12) = 9$$
$$3x + 6x - 36 = 9$$
$$9x - 36 = 9$$
$$9x = 45$$
$$x = 5$$
$$y = 2(5) - 12 = -2$$
The solution is $(5,-2)$.

33. Substituting into the first equation:
$$7x - 11(10) = 16$$
$$7x - 110 = 16$$
$$7x = 126$$
$$x = 18$$
$$y = 10$$
The solution is $(18,10)$.

35. Substituting into the first equation:
$$-4x + 4(x - 2) = -8$$
$$-4x + 4x - 8 = -8$$
$$-8 = -8$$
Since this statement is true, the system is dependent. The two lines coincide.

37. Substituting into the first equation:
$$0.05x + 0.10(22 - x) = 1.70$$
$$0.05x + 2.2 - 0.10x = 1.70$$
$$-0.05x + 2.2 = 1.7$$
$$-0.05x = -0.5$$
$$x = 10$$
$$y = 22 - 10 = 12$$
The solution is $(10,12)$.

39. **a.** At 1,000 miles the car and truck cost the same to operate.
    b. If Daniel drives more than 1,200 miles, the car will be cheaper to operate.
    c. If Daniel drives less than 800 miles, the truck will be cheaper to operate.
    d. The graphs appear in the first quadrant only because all quantities are positive.

41. Let x and $5x + 8$ represent the two numbers. The equation is:
$$x + 5x + 8 = 26$$
$$6x + 8 = 26$$
$$6x = 18$$
$$x = 3$$
$$5x + 8 = 23$$
The two numbers are 3 and 23.

43. Let x represent the smaller number and $2x - 6$ represent the larger number. The equation is:
$$2x - 6 - x = 9$$
$$x - 6 = 9$$
$$x = 15$$
$$2x - 6 = 24$$
The two numbers are 15 and 24.

45. Let w represent the width and $3w + 5$ represent the length. Using the perimeter formula:
$$2(w) + 2(3w + 5) = 58$$
$$2w + 6w + 10 = 58$$
$$8w + 10 = 58$$
$$8w = 48$$
$$w = 6$$
$$3w + 5 = 23$$
The width is 6 inches and the length is 23 inches.

47. Completing the table:

	Nickels	Dimes
Number	$x + 4$	x
Value (cents)	$5(x + 4)$	$10x$

The equation is:
$$5(x + 4) + 10x = 170$$
$$5x + 20 + 10x = 170$$
$$15x + 20 = 170$$
$$15x = 150$$
$$x = 10$$
$$x + 4 = 14$$
John has 14 nickels and 10 dimes.

7.4 Applications

1. Let x and y represent the two numbers. The system of equations is:
$$x + y = 25$$
$$y = 5 + x$$
Substituting into the first equation:
$$x + (5 + x) = 25$$
$$2x + 5 = 25$$
$$2x = 20$$
$$x = 10$$
$$y = 5 + 10 = 15$$
The two numbers are 10 and 15.

3. Let x and y represent the two numbers. The system of equations is:
$$x + y = 15$$
$$y = 4x$$
Substituting into the first equation:
$$x + 4x = 15$$
$$5x = 15$$
$$x = 3$$
$$y = 4(3) = 12$$
The two numbers are 3 and 12.

5.   Let x represent the larger number and y represent the smaller number. The system of equations is:
$$x - y = 5$$
$$x = 2y + 1$$
Substituting into the first equation:
$$2y + 1 - y = 5$$
$$y + 1 = 5$$
$$y = 4$$
$$x = 2(4) + 1 = 9$$
The two numbers are 4 and 9.

7.   Let x and y represent the two numbers. The system of equations is:
$$y = 4x + 5$$
$$x + y = 35$$
Substituting into the second equation:
$$x + 4x + 5 = 35$$
$$5x + 5 = 35$$
$$5x = 30$$
$$x = 6$$
$$y = 4(6) + 5 = 29$$
The two numbers are 6 and 29.

9.   Let x represent the amount invested at 6% and y represent the amount invested at 8%.
The system of equations is:
$$x + y = 20000$$
$$0.06x + 0.08y = 1380$$
Multiplying the first equation by -0.06:
$$-0.06x - 0.06y = -1200$$
$$0.06x + 0.08y = 1380$$
Adding the two equations:
$$0.02y = 180$$
$$y = 9000$$
Substituting into the first equation:
$$x + 9000 = 20000$$
$$x = 11000$$
Mr. Wilson invested $9,000 at 8% and $11,000 at 6%.

11.   Let x represent the amount invested at 5% and y represent the amount invested at 6%.
The system of equations is:
$$x = 4y$$
$$0.05x + 0.06y = 520$$
Substituting into the second equation:
$$0.05(4y) + 0.06y = 520$$
$$0.20y + 0.06y = 520$$
$$0.26y = 520$$
$$y = 2000$$
$$x = 4(2000) = 8000$$
She invested $8,000 at 5% and $2,000 at 6%.

13. Let x represent the number of nickels and y represent the number of quarters.
The system of equations is:
$$x + y = 14$$
$$0.05x + 0.25y = 2.30$$
Multiplying the first equation by -0.05:
$$-0.05x - 0.05y = -0.7$$
$$0.05x + 0.25y = 2.30$$
Adding the two equations:
$$0.20y = 1.6$$
$$y = 8$$
Substituting into the first equation:
$$x + 8 = 14$$
$$x = 6$$
Ron has 6 nickels and 8 quarters.

15. Let x represent the number of dimes and y represent the number of quarters.
The system of equations is:
$$x + y = 21$$
$$0.10x + 0.25y = 3.45$$
Multiplying the first equation by -0.10:
$$-0.10x - 0.10y = -2.10$$
$$0.10x + 0.25y = 3.45$$
Adding the two equations:
$$0.15y = 1.35$$
$$y = 9$$
Substituting into the first equation:
$$x + 9 = 21$$
$$x = 12$$
Tom has 12 dimes and 9 quarters.

17. Let x represent the liters of 50% alcohol solution and y represent the liters of 20% alcohol solution.
The system of equations is:
$$x + y = 18$$
$$0.50x + 0.20y = 0.30(18)$$
Multiplying the first equation by -0.20:
$$-0.20x - 0.20y = -3.6$$
$$0.50x + 0.20y = 5.4$$
Adding the two equations:
$$0.30x = 1.8$$
$$x = 6$$
Substituting into the first equation:
$$6 + y = 18$$
$$y = 12$$
The mixture contains 6 liters of 50% alcohol solution and 12 liters of 20% alcohol solution.

19. Let x represent the gallons of 10% disinfectant and y represent the gallons of 7% disinfectant. The system of equations is:
$$x + y = 30$$
$$0.10x + 0.07y = 0.08(30)$$
Multiplying the first equation by -0.07:
$$-0.07x - 0.07y = -2.1$$
$$0.10x + 0.07y = 2.4$$
Adding the two equations:
$$0.03x = 0.3$$
$$x = 10$$
Substituting into the first equation:
$$10 + y = 30$$
$$y = 20$$
The mixture contains 10 gallons of 10% disinfectant and 20 gallons of 7% disinfectant.

21. Let x represent the number of adult tickets and y represent the number of kids tickets. The system of equations is:
$$x + y = 70$$
$$5.50x + 4.00y = 310$$
Multiplying the first equation by -4:
$$-4.00x - 4.00y = -280$$
$$5.50x + 4.00y = 310$$
Adding the two equations:
$$1.5x = 30$$
$$x = 20$$
Substituting into the first equation:
$$20 + y = 70$$
$$y = 50$$
The matinee had 20 adult tickets sold and 50 kids tickets sold.

23. Let x represent the width and y represent the length. The system of equations is:
$$2x + 2y = 96$$
$$y = 2x$$
Substituting into the first equation:
$$2x + 2(2x) = 96$$
$$2x + 4x = 96$$
$$6x = 96$$
$$x = 16$$
$$y = 2(16) = 32$$
The width is 16 feet and the length is 32 feet.

25. Let x represent the number of $5 chips and y represent the number of $25 chips.
The system of equations is:
$$x + y = 45$$
$$5x + 25y = 465$$
Multiplying the first equation by -5:
$$-5x - 5y = -225$$
$$5x + 25y = 465$$
Adding the two equations:
$$20y = 240$$
$$y = 12$$
Substituting into the first equation:
$$x + 12 = 45$$
$$x = 33$$
The gambler has 33 $5 chips and 12 $25 chips.

27. Let x represent the number of shares of $11 stock and y represent the number of shares of $20 stock. The system of equations is:
$$x + y = 150$$
$$11x + 20y = 2550$$
Multiplying the first equation by -11:
$$-11x - 11y = -1650$$
$$11x + 20y = 2550$$
Adding the two equations:
$$9y = 900$$
$$y = 100$$
Substituting into the first equation:
$$x + 100 = 150$$
$$x = 50$$
She bought 50 shares at $11 and 100 shares at $20.

29. Reducing the rational expression: $\dfrac{x^2 - x - 6}{x^2 - 9} = \dfrac{(x-3)(x+2)}{(x+3)(x-3)} = \dfrac{x+2}{x+3}$

31. Performing the operations:
$$\frac{x^2 - 25}{x + 4} \cdot \frac{2x + 8}{x^2 - 9x + 20} = \frac{(x+5)(x-5)}{x+4} \cdot \frac{2(x+4)}{(x-4)(x-5)} = \frac{2(x+5)(x-5)(x+4)}{(x+4)(x-4)(x-5)} = \frac{2(x+5)}{x-4}$$

33. Performing the operations: $\dfrac{x}{x^2 - 16} + \dfrac{4}{x^2 - 16} = \dfrac{x+4}{x^2 - 16} = \dfrac{x+4}{(x+4)(x-4)} = \dfrac{1}{x-4}$

35. Simplifying the complex fraction: $\dfrac{1 - \dfrac{25}{x^2}}{1 - \dfrac{8}{x} + \dfrac{15}{x^2}} \cdot \dfrac{x^2}{x^2} = \dfrac{x^2 - 25}{x^2 - 8x + 15} = \dfrac{(x+5)(x-5)}{(x-5)(x-3)} = \dfrac{x+5}{x-3}$

37. Multiplying each side of the equation by $x^2 - 9 = (x+3)(x-3)$:

$$(x+3)(x-3)\left(\frac{x}{x^2-9} - \frac{3}{x-3}\right) = (x+3)(x-3) \bullet \frac{1}{x+3}$$
$$x - 3(x+3) = x - 3$$
$$x - 3x - 9 = x - 3$$
$$-2x - 9 = x - 3$$
$$-3x = 6$$
$$x = -2$$

Since $x = -2$ checks in the original equation, the solution is $x = -2$.

39. Let t represent the time to fill the pool with both pipes open. The equation is:

$$\frac{1}{8} - \frac{1}{12} = \frac{1}{t}$$
$$24t\left(\frac{1}{8} - \frac{1}{12}\right) = 24t \bullet \frac{1}{t}$$
$$3t - 2t = 24$$
$$t = 24$$

It will take 24 hours to fill the pool with both pipes left open.

41. The variation equation is $y = Kx$. Finding K:

$$8 = K \bullet 12$$
$$K = \frac{2}{3}$$

So $y = \frac{2}{3}x$. Substituting $x = 36$: $y = \frac{2}{3}(36) = 24$

Chapter 7 Test

See www.mathtv.com for video solutions to all problems in this chapter test.

Chapter 8
Roots, Radicals and More Quadratic Equations

8.1 Definitions and Common Roots

1. Finding the root: $\sqrt{4} = 2$

3. Finding the root: $-\sqrt{4} = -2$

5. $\sqrt{-16}$ is not a real number

7. Finding the root: $-\sqrt{144} = -12$

9. Finding the root: $\sqrt{625} = 25$

11. $\sqrt{-25}$ is not a real number

13. Finding the root: $-\sqrt{64} = -8$

15. Finding the root: $-\sqrt{100} = -10$

17. Finding the root: $-\sqrt{1,225} = -35$

19. Finding the root: $\sqrt[3]{1} = 1$

21. Finding the root: $\sqrt[3]{-8} = -2$

23. Finding the root: $-\sqrt[3]{125} = -5$

25. Finding the root: $\sqrt[3]{-1} = -1$

27. Finding the root: $\sqrt[3]{-27} = -3$

29. Finding the root: $-\sqrt[4]{16} = -2$

31. Simplifying the expression: $\sqrt{y^2} = y$

33. Simplifying the expression: $\sqrt{25x^2} = 5x$

35. Simplifying the expression: $\sqrt{a^2b^2} = ab$

37. Simplifying the expression: $\sqrt{(a+b)^2} = a+b$

39. Simplifying the expression: $\sqrt{81x^2y^2} = 9xy$

41. Simplifying the expression: $\sqrt[3]{x^3} = x$

43. Simplifying the expression: $\sqrt[3]{8x^3} = 2x$

45. Simplifying the expression: $\sqrt{x^4} = x^2$

47. Simplifying the expression: $\sqrt{36a^6} = 6a^3$

49. Simplifying the expression: $\sqrt{25a^8b^4} = 5a^4b^2$

51. **a.** Using a calculator: $\sqrt{18} \approx 4.243$ **b.** Using a calculator: $3\sqrt{2} \approx 4.243$

53. **a.** Using a calculator: $\sqrt{50} \approx 7.071$ **b.** Using a calculator: $5\sqrt{2} \approx 7.071$

55. **a.** Factoring the greatest common factor: $8 + 6x = 2(4 + 3x)$

    b. Factoring the greatest common factor: $8 + 6\sqrt{3} = 2\left(4 + 3\sqrt{3}\right)$

57. Simplifying the expression: $\sqrt{9} + \sqrt{16} = 3 + 4 = 7$

59. Simplifying the expression: $\sqrt{9+16} = \sqrt{25} = 5$

61. Simplifying the expression: $\sqrt{144} + \sqrt{25} = 12 + 5 = 17$

63. Simplifying the expression: $\sqrt{144+25} = \sqrt{169} = 13$

65. **a.** Approximating the expression: $\dfrac{1+\sqrt{5}}{2} \approx \dfrac{1+2.236}{2} = \dfrac{3.236}{2} = 1.618$

    b. Approximating the expression: $\dfrac{1-\sqrt{5}}{2} \approx \dfrac{1-2.236}{2} = \dfrac{-1.236}{2} = -0.618$

    c. Approximating the expression: $\dfrac{1+\sqrt{5}}{2} + \dfrac{1-\sqrt{5}}{2} \approx 1.618 - 0.618 = 1$

67. **a.**   Evaluating the root: $\sqrt{9} = 3$ **b.**   Evaluating the root: $\sqrt{900} = 30$

   c.   Evaluating the root: $\sqrt{0.09} = 0.3$

69. Simplifying each expression:

   $$\frac{5+\sqrt{49}}{2} = \frac{5+7}{2} = \frac{12}{2} = 6 \qquad \frac{5-\sqrt{49}}{2} = \frac{5-7}{2} = \frac{-2}{2} = -1$$

71. **a.**   Simplifying the expression: $\dfrac{2\sqrt{3}}{10} = \dfrac{\sqrt{3}}{5}$

   b.   Simplifying the expression: $\dfrac{5\sqrt{2}}{15} = \dfrac{\sqrt{2}}{3}$

   c.   Simplifying the expression: $\dfrac{15+6\sqrt{3}}{12} = \dfrac{3\left(5+2\sqrt{3}\right)}{12} = \dfrac{5+2\sqrt{3}}{4}$

   d.   Simplifying the expression: $\dfrac{5+10\sqrt{6}}{5} = \dfrac{5\left(1+2\sqrt{6}\right)}{5} = 1+2\sqrt{6}$

73. Using the Pythagorean Theorem: 75. Using the Pythagorean Theorem:

   $$x^2 = 3^2 + 4^2 = 9 + 16 = 25 \qquad\qquad x^2 = 5^2 + 10^2 = 25 + 100 = 125$$

   $$x = \sqrt{25} = 5 \qquad\qquad\qquad\qquad\quad x = \sqrt{125} \approx 11.2$$

77. Let l represent the length of the wire. Using the Pythagorean Theorem:

   $$l^2 = 24^2 + 18^2 = 576 + 324 = 900$$

   $$l = \sqrt{900} = 30$$

   The length of the wire is 30 feet.

79. Using a calculator: $\dfrac{1+\sqrt{5}}{2} \approx 1.618$ 81. Simplifying: $3 \bullet \sqrt{16} = 3 \bullet 4 = 12$

83. Factoring: $75 = 25 \bullet 3$ 85. Factoring: $50 = 25 \bullet 2$

87. Factoring: $40 = 4 \bullet 10$ 89. Factoring: $x^2 = x^2$

91. Factoring: $12x^2 = 4x^2 \bullet 3$ 93. Factoring: $50x^3 y^2 = 25x^2 y^2 \bullet 2x$

8.2 Properties of Radicals

1.   Simplifying the radical expression: $\sqrt{8} = \sqrt{4 \bullet 2} = \sqrt{4}\sqrt{2} = 2\sqrt{2}$

3.   Simplifying the radical expression: $\sqrt{12} = \sqrt{4 \bullet 3} = \sqrt{4}\sqrt{3} = 2\sqrt{3}$

5.   Simplifying the radical expression: $\sqrt[3]{24} = \sqrt[3]{8 \bullet 3} = \sqrt[3]{8}\sqrt[3]{3} = 2\sqrt[3]{3}$

7.   Simplifying the radical expression: $\sqrt{50x^2} = \sqrt{25x^2 \bullet 2} = \sqrt{25x^2}\sqrt{2} = 5x\sqrt{2}$

9.   Simplifying the radical expression: $\sqrt{45a^2b^2} = \sqrt{9a^2b^2 \bullet 5} = \sqrt{9a^2b^2}\sqrt{5} = 3ab\sqrt{5}$

11. Simplifying the radical expression: $\sqrt[3]{54x^3} = \sqrt[3]{27x^3 \bullet 2} = \sqrt[3]{27x^3}\sqrt[3]{2} = 3x\sqrt[3]{2}$

13. Simplifying the radical expression: $\sqrt{32x^4} = \sqrt{16x^4 \bullet 2} = \sqrt{16x^4}\sqrt{2} = 4x^2\sqrt{2}$

15. Simplifying the radical expression: $5\sqrt{80} = 5\sqrt{16 \bullet 5} = 5\sqrt{16}\sqrt{5} = 5 \bullet 4\sqrt{5} = 20\sqrt{5}$

17. Simplifying the radical expression: $\dfrac{1}{2}\sqrt{28x^3} = \dfrac{1}{2}\sqrt{4x^2 \bullet 7x} = \dfrac{1}{2}\sqrt{4x^2}\sqrt{7x} = \dfrac{1}{2} \bullet 2x\sqrt{7x} = x\sqrt{7x}$

19. Simplifying the radical expression: $x\sqrt[3]{2x^4} = x\sqrt[3]{x^3 \bullet 2x} = x\sqrt[3]{x^3}\sqrt[3]{2x} = x \bullet x\sqrt[3]{x}2 = x^2\sqrt[3]{2x}$

21. Simplifying the radical expression:

$$2a\sqrt[3]{27a^5} = 2a\sqrt[3]{27a^3 \cdot a^2} = 2a\sqrt[3]{27a^3}\sqrt[3]{a^2} = 2a \cdot 3a\sqrt[3]{a^2} = 6a^2\sqrt[3]{a^2}$$

23. Simplifying the radical expression:

$$\frac{4}{3}\sqrt{45a^3} = \frac{4}{3}\sqrt{9a^2 \cdot 5a} = \frac{4}{3}\sqrt{9a^2}\sqrt{5a} = \frac{4}{3} \cdot 3a\sqrt{5a} = 4a\sqrt{5a}$$

25. Simplifying the radical expression:

$$3\sqrt{50xy^2} = 3\sqrt{25y^2 \cdot 2x} = 3\sqrt{25y^2}\sqrt{2x} = 3 \cdot 5y\sqrt{2x} = 15y\sqrt{2x}$$

27. Simplifying the radical expression:

$$7\sqrt{12x^2y} = 7\sqrt{4x^2 \cdot 3y} = 7\sqrt{4x^2}\sqrt{3y} = 7 \cdot 2x\sqrt{3y} = 14x\sqrt{3y}$$

29. Simplifying the radical expression: $\sqrt{\dfrac{16}{25}} = \dfrac{\sqrt{16}}{\sqrt{25}} = \dfrac{4}{5}$

31. Simplifying the radical expression: $\sqrt{\dfrac{4}{9}} = \dfrac{\sqrt{4}}{\sqrt{9}} = \dfrac{2}{3}$

33. Simplifying the radical expression: $\sqrt[3]{\dfrac{8}{27}} = \dfrac{\sqrt[3]{8}}{\sqrt[3]{27}} = \dfrac{2}{3}$

35. Simplifying the radical expression: $\sqrt[4]{\dfrac{16}{81}} = \dfrac{\sqrt[4]{16}}{\sqrt[4]{81}} = \dfrac{2}{3}$

37. Simplifying the radical expression: $\sqrt{\dfrac{100x^2}{25}} = \sqrt{4x^2} = 2x$

39. Simplifying the radical expression: $\sqrt{\dfrac{81a^2b^2}{9}} = \sqrt{9a^2b^2} = 3ab$

41. Simplifying the radical expression: $\sqrt[3]{\dfrac{27x^3}{8y^3}} = \dfrac{\sqrt[3]{27x^3}}{\sqrt[3]{8y^3}} = \dfrac{3x}{2y}$

43. Simplifying the radical expression: $\sqrt{\dfrac{50}{9}} = \dfrac{\sqrt{50}}{\sqrt{9}} = \dfrac{\sqrt{25 \cdot 2}}{3} = \dfrac{5\sqrt{2}}{3}$

45. Simplifying the radical expression: $\sqrt{\dfrac{75}{25}} = \sqrt{3}$

47. Simplifying the radical expression: $\sqrt{\dfrac{128}{49}} = \dfrac{\sqrt{128}}{\sqrt{49}} = \dfrac{\sqrt{64 \cdot 2}}{7} = \dfrac{8\sqrt{2}}{7}$

49. Simplifying the radical expression: $\sqrt{\dfrac{288x}{25}} = \dfrac{\sqrt{288x}}{\sqrt{25}} = \dfrac{\sqrt{144 \cdot 2x}}{5} = \dfrac{12\sqrt{2x}}{5}$

51. Simplifying the radical expression: $\sqrt{\dfrac{54a^2}{25}} = \dfrac{\sqrt{54a^2}}{\sqrt{25}} = \dfrac{\sqrt{9a^2 \cdot 6}}{5} = \dfrac{3a\sqrt{6}}{5}$

53. Simplifying the radical expression: $\dfrac{3\sqrt{50}}{2} = \dfrac{3\sqrt{25 \cdot 2}}{2} = \dfrac{3 \cdot 5\sqrt{2}}{2} = \dfrac{15\sqrt{2}}{2}$

55. Simplifying the radical expression: $\dfrac{7\sqrt{28y^2}}{3} = \dfrac{7\sqrt{4y^2 \cdot 7}}{3} = \dfrac{7 \cdot 2y\sqrt{7}}{3} = \dfrac{14y\sqrt{7}}{3}$

57. Simplifying the radical expression:

$$\frac{5\sqrt{72a^2b^2}}{\sqrt{36}} = \frac{5\sqrt{36a^2b^2 \bullet 2}}{6} = \frac{5 \bullet 6ab\sqrt{2}}{6} = \frac{30ab\sqrt{2}}{6} = 5ab\sqrt{2}$$

59. **a.** Simplifying the expression: $\dfrac{\sqrt{20}}{4} = \dfrac{\sqrt{4 \bullet 5}}{4} = \dfrac{2\sqrt{5}}{4} = \dfrac{\sqrt{5}}{2}$

  b. Simplifying the expression: $\dfrac{3\sqrt{20}}{15} = \dfrac{3\sqrt{4 \bullet 5}}{15} = \dfrac{6\sqrt{5}}{15} = \dfrac{2\sqrt{5}}{5}$

  c. Simplifying the expression: $\dfrac{4+\sqrt{12}}{2} = \dfrac{4+2\sqrt{3}}{2} = \dfrac{2\left(2+\sqrt{3}\right)}{2} = 2+\sqrt{3}$

  d. Simplifying the expression: $\dfrac{2+\sqrt{9}}{5} = \dfrac{2+3}{5} = \dfrac{5}{5} = 1$

61. **a.** Substituting the values: $\sqrt{b^2-4ac} = \sqrt{(4)^2 - 4(2)(-3)} = \sqrt{16+24} = \sqrt{40} = \sqrt{4 \bullet 10} = 2\sqrt{10}$

  b. Substituting the values: $\sqrt{b^2-4ac} = \sqrt{(1)^2 - 4(1)(-6)} = \sqrt{1+24} = \sqrt{25} = 5$

  c. Substituting the values: $\sqrt{b^2-4ac} = \sqrt{(1)^2 - 4(1)(-11)} = \sqrt{1+44} = \sqrt{45} = \sqrt{9 \bullet 5} = 3\sqrt{5}$

  d. Substituting the values: $\sqrt{b^2-4ac} = \sqrt{(6)^2 - 4(3)(2)} = \sqrt{36-24} = \sqrt{12} = \sqrt{4 \bullet 3} = 2\sqrt{3}$

63. **a.** Simplifying: $\sqrt{32x^{10}y^5} = \sqrt{16x^{10}y^4 \bullet 2y} = \sqrt{16x^{10}y^4}\sqrt{2y} = 4x^5y^2\sqrt{2y}$

  b. Simplifying: $\sqrt[3]{32x^{10}y^5} = \sqrt[3]{8x^9y^3 \bullet 4xy^2} = \sqrt[3]{8x^9y^3}\sqrt[3]{4xy^2} = 2x^3y\sqrt[3]{4xy^2}$

  c. Simplifying: $\sqrt[4]{32x^{10}y^5} = \sqrt[4]{16x^8y^4 \bullet 2x^2y} = \sqrt[4]{16x^8y^4}\sqrt[4]{2x^2y} = 2x^2y\sqrt[4]{2x^2y}$

  d. Simplifying: $\sqrt[5]{32x^{10}y^5} = 2x^2y$

65. **a.** Simplifying: $\sqrt{4} = 2$ 　　　　　**b.** Simplifying: $\sqrt{0.04} = 0.2$

  c. Simplifying: $\sqrt{400} = 20$ 　　　**d.** Simplifying: $\sqrt{0.0004} = 0.02$

67. Completing the table:

x	$\sqrt{x}$	$2\sqrt{x}$	$\sqrt{4x}$
1	1	2	2
2	1.414	2.828	2.828
3	1.732	3.464	3.464
4	2	4	4

69. Completing the table:

x	$\sqrt{x}$	$3\sqrt{x}$	$\sqrt{9x}$
1	1	3	3
2	1.414	4.243	4.243
3	1.732	5.196	5.196
4	2	6	6

71. Substituting $h = 25$: $t = \sqrt{\dfrac{25}{16}} = \dfrac{5}{4}$ seconds

73. Simplifying: $\sqrt{4x^3y^2} = \sqrt{4x^2y^2 \bullet x} = 2xy\sqrt{x}$

75. Simplifying: $\dfrac{6}{2}\sqrt{16} = 3 \bullet 4 = 12$

77. Simplifying: $\dfrac{\sqrt{2}}{\sqrt{4}} = \dfrac{\sqrt{2}}{2}$

79. Simplifying: $\dfrac{\sqrt[3]{18}}{\sqrt[3]{9}} = \sqrt[3]{\dfrac{18}{9}} = \sqrt[3]{2}$

81. Multiplying: $\dfrac{\sqrt{2}}{\sqrt{3}} \bullet \dfrac{\sqrt{3}}{\sqrt{3}} = \dfrac{\sqrt{6}}{\sqrt{9}} = \dfrac{\sqrt{6}}{3}$

83. Multiplying: $\sqrt[3]{3} \bullet \sqrt[3]{9} = \sqrt[3]{27} = 3$

8.3 Simplified Form for Radicals

1. Simplifying the radical expression: $\sqrt{\dfrac{1}{2}} = \dfrac{\sqrt{1}}{\sqrt{2}} \cdot \dfrac{\sqrt{2}}{\sqrt{2}} = \dfrac{\sqrt{2}}{\sqrt{4}} = \dfrac{\sqrt{2}}{2}$

3. Simplifying the radical expression: $\sqrt{\dfrac{1}{3}} = \dfrac{\sqrt{1}}{\sqrt{3}} \cdot \dfrac{\sqrt{3}}{\sqrt{3}} = \dfrac{\sqrt{3}}{\sqrt{9}} = \dfrac{\sqrt{3}}{3}$

5. Simplifying the radical expression: $\sqrt{\dfrac{2}{5}} = \dfrac{\sqrt{2}}{\sqrt{5}} \cdot \dfrac{\sqrt{5}}{\sqrt{5}} = \dfrac{\sqrt{10}}{\sqrt{25}} = \dfrac{\sqrt{10}}{5}$

7. Simplifying the radical expression: $\sqrt{\dfrac{3}{2}} = \dfrac{\sqrt{3}}{\sqrt{2}} \cdot \dfrac{\sqrt{2}}{\sqrt{2}} = \dfrac{\sqrt{6}}{\sqrt{4}} = \dfrac{\sqrt{6}}{2}$

9. Simplifying the radical expression: $\sqrt{\dfrac{20}{2}} = \sqrt{10}$

11. Simplifying the radical expression: $\sqrt{\dfrac{45}{6}} = \sqrt{\dfrac{15}{2}} = \dfrac{\sqrt{15}}{\sqrt{2}} \cdot \dfrac{\sqrt{2}}{\sqrt{2}} = \dfrac{\sqrt{30}}{\sqrt{4}} = \dfrac{\sqrt{30}}{2}$

13. Simplifying the radical expression: $\sqrt{\dfrac{20}{5}} = \sqrt{4} = 2$

15. Simplifying the radical expression: $\dfrac{\sqrt{21}}{\sqrt{3}} = \sqrt{\dfrac{21}{3}} = \sqrt{7}$

17. Simplifying the radical expression: $\dfrac{\sqrt{35}}{\sqrt{7}} = \sqrt{\dfrac{35}{7}} = \sqrt{5}$

19. Simplifying the radical expression: $\dfrac{10\sqrt{15}}{5\sqrt{3}} = \dfrac{10}{5} \cdot \sqrt{\dfrac{15}{3}} = 2\sqrt{5}$

21. Simplifying the radical expression: $\dfrac{6\sqrt{21}}{3\sqrt{7}} = \dfrac{6}{3} \cdot \sqrt{\dfrac{21}{7}} = 2\sqrt{3}$

23. Simplifying the radical expression: $\dfrac{6\sqrt{35}}{12\sqrt{5}} = \dfrac{6}{12} \cdot \sqrt{\dfrac{35}{5}} = \dfrac{1}{2}\sqrt{7} = \dfrac{\sqrt{7}}{2}$

25. Simplifying the radical expression: $\sqrt{\dfrac{4x^2y^2}{2}} = \sqrt{2x^2y^2} = xy\sqrt{2}$

27. Simplifying the radical expression: $\sqrt{\dfrac{5x^2y}{3}} = \dfrac{\sqrt{5x^2y}}{\sqrt{3}} \cdot \dfrac{\sqrt{3}}{\sqrt{3}} = \dfrac{\sqrt{15x^2y}}{\sqrt{9}} = \dfrac{x\sqrt{15y}}{3}$

29. Simplifying the radical expression: $\sqrt{\dfrac{16a^4}{5}} = \dfrac{\sqrt{16a^4}}{\sqrt{5}} \cdot \dfrac{\sqrt{5}}{\sqrt{5}} = \dfrac{4a^2\sqrt{5}}{\sqrt{25}} = \dfrac{4a^2\sqrt{5}}{5}$

31. Simplifying the radical expression:

$$\sqrt{\dfrac{72a^5}{5}} = \dfrac{\sqrt{72a^5}}{\sqrt{5}} \cdot \dfrac{\sqrt{5}}{\sqrt{5}} = \dfrac{\sqrt{360a^5}}{\sqrt{25}} = \dfrac{\sqrt{36a^4 \cdot 10a}}{5} = \dfrac{6a^2\sqrt{10a}}{5}$$

33.   Simplifying the radical expression:

$$\sqrt{\frac{20x^2y^3}{3}} = \frac{\sqrt{20x^2y^3}}{\sqrt{3}} \cdot \frac{\sqrt{3}}{\sqrt{3}} = \frac{\sqrt{60x^2y^3}}{\sqrt{9}} = \frac{\sqrt{4x^2y^2 \cdot 15y}}{3} = \frac{2xy\sqrt{15y}}{3}$$

35.   Simplifying the radical expression: $\dfrac{2\sqrt{20x^2y^3}}{3} = \dfrac{2\sqrt{4x^2y^2 \cdot 5y}}{3} = \dfrac{2 \cdot 2xy\sqrt{5y}}{3} = \dfrac{4xy\sqrt{5y}}{3}$

37.   Simplifying the radical expression: $\dfrac{6\sqrt{54a^2b^3}}{5} = \dfrac{6\sqrt{9a^2b^2 \cdot 6b}}{5} = \dfrac{6 \cdot 3ab\sqrt{6b}}{5} = \dfrac{18ab\sqrt{6b}}{5}$

39.   a.   Simplifying the expression: $\dfrac{10+\sqrt{75}}{5} = \dfrac{10+5\sqrt{3}}{5} = \dfrac{5\left(2+\sqrt{3}\right)}{5} = 2+\sqrt{3}$

  b.   Simplifying the expression: $\dfrac{-6+\sqrt{45}}{3} = \dfrac{-6+3\sqrt{5}}{3} = \dfrac{3\left(-2+\sqrt{5}\right)}{3} = -2+\sqrt{5}$

  c.   Simplifying the expression: $\dfrac{-2-\sqrt{27}}{6} = \dfrac{-2-3\sqrt{3}}{6}$

41.   Completing the table:

x	$\sqrt{x}$	$\dfrac{1}{\sqrt{x}}$	$\dfrac{\sqrt{x}}{x}$
1	1	1	1
2	1.414	0.707	0.707
3	1.732	0.577	0.577
4	2	0.5	0.5
5	2.236	0.447	0.447
6	2.449	0.408	0.408

43.   Completing the table:

x	$\sqrt{x^2}$	$\sqrt{x^3}$	$x\sqrt{x}$
1	1	1	1
2	2	2.828	2.828
3	3	5.196	5.196
4	4	8	8
5	5	11.180	11.180
6	6	14.697	14.697

45.   Substituting $h = 24$: $d = \sqrt{\dfrac{3(24)}{2}} = \sqrt{\dfrac{72}{2}} = \sqrt{36} = 6$ miles

47.   Drawing the figure:

49. Simplifying the terms in the sequence:
$$\sqrt{1^2 + 1} = \sqrt{1+1} = \sqrt{2}$$
$$\sqrt{\left(\sqrt{2}\right)^2 + 1} = \sqrt{2+1} = \sqrt{3}$$
$$\sqrt{\left(\sqrt{3}\right)^2 + 1} = \sqrt{3+1} = \sqrt{4} = 2$$

51. Factoring: $x^2 - 16 = (x+4)(x-4)$

53. Factoring: $16a^2 - 25 = (4a+5)(4a-5)$

55. Factoring: $2x^2 - 5x - 12 = (2x+3)(x-4)$

57. Factoring: $2x^2 + 11x + 12 = (2x+3)(x+4)$

59. Solving the equation:
$$y + 4 = 5$$
$$y = 1$$

61. Solving the equation:
$$2x + 3 = 0$$
$$2x = -3$$
$$x = -\frac{3}{2}$$

8.4 More Quadratic Equations

1. Solving by factoring:
$$x^2 - 25 = 0$$
$$(x+5)(x-5) = 0$$
$$x = -5, 5$$

3. Solving by factoring:
$$4x^2 - 9 = 0$$
$$(2x+3)(2x-3) = 0$$
$$x = -\frac{3}{2}, \frac{3}{2}$$

5. Solving by factoring:
$$2x^2 + 3x - 5 = 0$$
$$(2x+5)(x-1) = 0$$
$$x = -\frac{5}{2}, 1$$

7. Solving by factoring:
$$a^2 - a - 6 = 0$$
$$(a+2)(a-3) = 0$$
$$a = -2, 3$$

9. Solving the equation:
$$x^2 = 25$$
$$x = \pm\sqrt{25}$$
$$x = \pm 5$$

11. Solving the equation:
$$4x^2 = 9$$
$$x^2 = \frac{9}{4}$$
$$x = \pm\sqrt{\frac{9}{4}}$$
$$x = \pm\frac{3}{2}$$

13. Solving the equation:
$$a^2 = 8$$
$$a = \pm\sqrt{8}$$
$$a = \pm 2\sqrt{2}$$

15. Solving the equation:
$$2x^2 = 24$$
$$x^2 = 12$$
$$x = \pm\sqrt{12}$$
$$x = \pm 2\sqrt{3}$$

17. Solving the equation:
$$(x+2)^2 = 4$$
$$x+2 = \pm\sqrt{4}$$
$$x+2 = \pm 2$$
$$x = -2 \pm 2$$
$$x = -4, 0$$

19. Solving the equation:
$$(x+1)^2 = 16$$
$$x+1 = \pm\sqrt{16}$$
$$x+1 = \pm 4$$
$$x = -1 \pm 4$$
$$x = -5, 3$$

21. Solving the equation:
$$(a-5)^2 = 75$$
$$a-5 = \pm\sqrt{75}$$
$$a-5 = \pm 5\sqrt{3}$$
$$a = 5 \pm 5\sqrt{3}$$

23. Solving the equation:
$$(y+1)^2 = 12$$
$$y+1 = \pm\sqrt{12}$$
$$y+1 = \pm 2\sqrt{2}$$
$$y = -1 \pm 2\sqrt{3}$$

25. Solving the equation:
$$(2x+1)^2 = 25$$
$$2x+1 = \pm\sqrt{25}$$
$$2x+1 = \pm 5$$
$$2x = -1 \pm 5$$
$$2x = -6, 4$$
$$x = -3, 2$$

27. Solving the equation:
$$(4a-5)^2 = 36$$
$$4a-5 = \pm\sqrt{36}$$
$$4a-5 = \pm 6$$
$$4a = 5 \pm 6$$
$$4a = -1, 11$$
$$a = -\frac{1}{4}, \frac{11}{4}$$

29. Solving the equation:
$$(3y-1)^2 = 20$$
$$3y-1 = \pm\sqrt{20}$$
$$3y-1 = \pm 2\sqrt{5}$$
$$3y = 1 \pm 2\sqrt{5}$$
$$y = \frac{1 \pm 2\sqrt{5}}{3}$$

31. Solving the equation:
$$(3x+6)^2 = 27$$
$$3x+6 = \pm\sqrt{27}$$
$$3x+6 = \pm 3\sqrt{3}$$
$$3x = -6 \pm 3\sqrt{3}$$
$$x = -2 \pm \sqrt{3}$$

33. Solving the equation:
$$(3x-9)^2 = 27$$
$$3x-9 = \pm\sqrt{27}$$
$$3x-9 = \pm 3\sqrt{3}$$
$$3x = 9 \pm 3\sqrt{3}$$
$$x = 3 \pm \sqrt{3}$$

35. Solving the equation:
$$\left(x-\frac{1}{2}\right)^2 = \frac{7}{4}$$
$$x-\frac{1}{2} = \pm\sqrt{\frac{7}{4}}$$
$$x-\frac{1}{2} = \pm\frac{\sqrt{7}}{2}$$
$$x = \frac{1 \pm \sqrt{7}}{2}$$

37. Solving the equation:
$$\left(a+\frac{4}{5}\right)^2 = \frac{12}{25}$$
$$a+\frac{4}{5} = \pm\sqrt{\frac{12}{25}}$$
$$a+\frac{4}{5} = \pm\frac{2\sqrt{3}}{5}$$
$$a = \frac{-4\pm2\sqrt{3}}{5}$$

39. Solving the equation:
$$x^2 + 10x + 25 = 7$$
$$(x+5)^2 = 7$$
$$x+5 = \pm\sqrt{7}$$
$$x = -5 \pm \sqrt{7}$$

41. Solving the equation:
$$x^2 - 2x + 1 = 9$$
$$(x-1)^2 = 9$$
$$x-1 = \pm\sqrt{9}$$
$$x-1 = \pm3$$
$$x = 1 \pm 3$$
$$x = -2, 4$$

43. Solving the equation:
$$x^2 + 12x + 36 = 8$$
$$(x+6)^2 = 8$$
$$x+6 = \pm\sqrt{8}$$
$$x+6 = \pm2\sqrt{2}$$
$$x = -6 \pm 2\sqrt{2}$$

45. a. Solving the equation:
$$2x - 1 = 0$$
$$2x = 1$$
$$x = \frac{1}{2}$$

b. Solving the equation:
$$2x - 1 = 4$$
$$2x = 5$$
$$x = \frac{5}{2}$$

c. Solving the equation:
$$(2x-1)^2 = 4$$
$$2x-1 = \pm\sqrt{4}$$
$$2x-1 = \pm2$$
$$2x = 1 \pm 2$$
$$2x = -1, 3$$
$$x = -\frac{1}{2}, \frac{3}{2}$$

d. Solving the equation:
$$\frac{1}{2x} - 1 = \frac{1}{4}$$
$$4x\left(\frac{1}{2x}-1\right) = 4x\left(\frac{1}{4}\right)$$
$$2 - 4x = x$$
$$2 = 5x$$
$$x = \frac{2}{5}$$

47. Checking the solution: $(x+1)^2 = \left(-1+5\sqrt{2}+1\right)^2 = \left(5\sqrt{2}\right)^2 = 25 \bullet 2 = 50$

49. Let x represent the number. The equation is:
$$(x+3)^2 = 16$$
$$x+3 = \pm\sqrt{16}$$
$$x+3 = \pm4$$
$$x = -3 \pm 4$$
$$x = -7, 1$$
The number is either -7 or 1.

51. Let x represent the horizontal distance traveled. Using the Pythagorean theorem:
$$x^2 + 120^2 = 790^2$$
$$x^2 + 14400 = 624100$$
$$x^2 + 14400 = 624100$$
$$x^2 = 609700$$
$$x = \sqrt{609700} \approx 781$$
The horizontal distance is approximately 781 feet.

53. Let x represent the height of the triangle. Draw the figure:

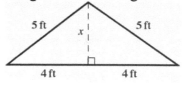

Using the Pythagorean theorem:
$$4^2 + x^2 = 5^2$$
$$16 + x^2 = 25$$
$$x^2 = 9$$
$$x = \sqrt{9} = 3$$
The height is 3 feet.

55. Simplifying: $\left(\dfrac{1}{2} \cdot 18\right)^2 = (9)^2 = 81$

57. Simplifying: $\left[\dfrac{1}{2}(-2)\right]^2 = (-1)^2 = 1$

59. Simplifying: $\left(\dfrac{1}{2} \cdot 3\right)^2 = \left(\dfrac{3}{2}\right)^2 = \dfrac{9}{4}$

61. Simplifying: $\dfrac{2x^2 + 16}{2} = \dfrac{2(x^2 + 8)}{2} = x^2 + 8$

63. Factoring: $x^2 + 6x + 9 = (x+3)^2$

65. Factoring: $y^2 - 3y + \dfrac{9}{4} = \left(y - \dfrac{3}{2}\right)^2$

8.5 Completing the Square

1. The correct term is 36, since: $x^2 + 12x + 36 = (x+6)^2$

3. The correct term is 36, since: $x^2 - 12x + 36 = (x-6)^2$

5. The correct term is 4, since: $x^2 + 4x + 4 = (x+2)^2$

7. The correct term is 4, since: $x^2 - 4x + 4 = (x-2)^2$

9. The correct term is 25, since: $a^2 - 10a + 25 = (a-5)^2$

11. The correct term is $\dfrac{25}{4}$, since: $x^2 + 5x + \dfrac{25}{4} = \left(x + \dfrac{5}{2}\right)^2$

13. The correct term is $\dfrac{1}{4}$, since: $y^2 + y + \dfrac{1}{4} = \left(y + \dfrac{1}{2}\right)^2$

15. The correct term is $\dfrac{1}{16}$, since: $x^2 + \dfrac{1}{2}x + \dfrac{1}{16} = \left(x + \dfrac{1}{4}\right)^2$

17. The correct term is $\dfrac{1}{9}$, since: $x^2 + \dfrac{2}{3}x + \dfrac{1}{9} = \left(x + \dfrac{1}{3}\right)^2$

19. Solve by completing the square:
$$x^2 + 4x = 12$$
$$x^2 + 4x + 4 = 12 + 4$$
$$(x+2)^2 = 16$$
$$x + 2 = \pm 4$$
$$x = -2 \pm 4$$
$$x = -6, 2$$

21. Solve by completing the square:
$$x^2 + 12x = -27$$
$$x^2 + 12x + 36 = -27 + 36$$
$$(x+6)^2 = 9$$
$$x + 6 = \pm 3$$
$$x = -6 \pm 3$$
$$x = -9, -3$$

23. Solve by completing the square:
$$a^2 - 2a - 5 = 0$$
$$a^2 - 2a = 5$$
$$a^2 - 2a + 1 = 5 + 1$$
$$(a-1)^2 = 6$$
$$a - 1 = \pm\sqrt{6}$$
$$a = 1 \pm \sqrt{6}$$

25. Solve by completing the square:
$$y^2 - 8y + 1 = 0$$
$$y^2 - 8y = -1$$
$$y^2 - 8y + 16 = -1 + 16$$
$$(y-4)^2 = 15$$
$$y - 4 = \pm\sqrt{15}$$
$$y = 4 \pm \sqrt{15}$$

27. Solve by completing the square:
$$x^2 - 5x - 3 = 0$$
$$x^2 - 5x = 3$$
$$x^2 - 5x + \frac{25}{4} = 3 + \frac{25}{4}$$
$$\left(x - \frac{5}{2}\right)^2 = \frac{37}{4}$$
$$x - \frac{5}{2} = \pm\sqrt{\frac{37}{4}} = \frac{\pm\sqrt{37}}{2}$$
$$x = \frac{5 \pm \sqrt{37}}{2}$$

29. Solve by completing the square:
$$2x^2 - 4x - 8 = 0$$
$$x^2 - 2x - 4 = 0$$
$$x^2 - 2x = 4$$
$$x^2 - 2x + 1 = 4 + 1$$
$$(x-1)^2 = 5$$
$$x - 1 = \pm\sqrt{5}$$
$$x = 1 \pm \sqrt{5}$$

31. Solve by completing the square:
$$x^2 - 10x = 0$$
$$x^2 - 10x + 25 = 0 + 25$$
$$(x-5)^2 = 25$$
$$x - 5 = \pm 5$$
$$x = 5 \pm 5$$
$$x = 0, 10$$

33. Solve by completing the square:
$$y^2 + 2y - 15 = 0$$
$$y^2 + 2y = 15$$
$$y^2 + 2y + 1 = 15 + 1$$
$$(y+1)^2 = 16$$
$$y + 1 = \pm 4$$
$$y = -1 \pm 4$$
$$y = -5, 3$$

35. Solve by completing the square:

$$x^2 + 6x = -5$$
$$x^2 + 6x + 9 = -5 + 9$$
$$(x+3)^2 = 4$$
$$x + 3 = \pm 2$$
$$x = -3 \pm 2$$
$$x = -5, -1$$

37. Solve by completing the square:

$$x^2 - 3x = -2$$
$$x^2 - 3x + \frac{9}{4} = -2 + \frac{9}{4}$$
$$\left(x - \frac{3}{2}\right)^2 = \frac{1}{4}$$
$$x - \frac{3}{2} = \pm \frac{1}{2}$$
$$x = \frac{3}{2} \pm \frac{1}{2}$$
$$x = 1, 2$$

39. Solve by completing the square:

$$4x^2 + 8x - 4 = 0$$
$$x^2 + 2x - 1 = 0$$
$$x^2 + 2x = 1$$
$$x^2 + 2x + 1 = 1 + 1$$
$$(x+1)^2 = 2$$
$$x + 1 = \pm\sqrt{2}$$
$$x = -1 \pm \sqrt{2}$$

41. Solve by completing the square:

$$2x^2 + 2x - 4 = 0$$
$$x^2 + x - 2 = 0$$
$$x^2 + x = 2$$
$$x^2 + x + \frac{1}{4} = 2 + \frac{1}{4}$$
$$\left(x + \frac{1}{2}\right)^2 = \frac{9}{4}$$
$$x + \frac{1}{2} = \pm \frac{3}{2}$$
$$x = -\frac{1}{2} \pm \frac{3}{2}$$
$$x = -2, 1$$

43. a. Solving by factoring:

$$x^2 - 6x = 0$$
$$x(x-6) = 0$$
$$x = 0, 6$$

b. Solving by completing the square:

$$x^2 - 6x = 0$$
$$x^2 - 6x + 9 = 9$$
$$(x-3)^2 = 9$$
$$x - 3 = \pm\sqrt{9}$$
$$x - 3 = -3, 3$$
$$x = 0, 6$$

45. a. Solving by factoring:

$$x^2 + 2x = 35$$
$$x^2 + 2x - 35 = 0$$
$$(x+7)(x-5) = 0$$
$$x = -7, 5$$

b. Solving by completing the square:

$$x^2 + 2x = 35$$
$$x^2 + 2x + 1 = 35 + 1$$
$$(x+1)^2 = 36$$
$$x + 1 = \pm 6$$
$$x + 1 = -6, 6$$
$$x = -7, 5$$

47. Using the Pythagorean theorem:
$$(3x)^2 + (4x)^2 = 14^2$$
$$9x^2 + 16x^2 = 196$$
$$25x^2 = 196$$
$$x^2 = \frac{196}{25}$$
$$x = \sqrt{\frac{196}{25}} = 2.8$$
$$3x = 8.4, 4x = 11.2$$
The length (height) of the screen is 8.4 inches and the width of the screen is 11.2 inches.

49. Solving the equation:
$$x^2 - 2x - 1 = 0$$
$$x^2 - 2x = 1$$
$$x^2 - 2x + 1 = 1 + 1$$
$$(x-1)^2 = 2$$
$$x - 1 = \pm\sqrt{2}$$
$$x = 1 \pm \sqrt{2} \approx -0.4, 2.4$$

51. **a.** Finding the sum: $\left(-2+\sqrt{7}\right)+\left(-2-\sqrt{7}\right) = -2+\sqrt{7}-2-\sqrt{7} = -4$

b. Finding the product: $\left(-2+\sqrt{7}\right)\left(-2-\sqrt{7}\right) = 4 - 2\sqrt{7} + 2\sqrt{7} - 7 = -3$

53. The square has an area of 9.   **55.** The square has an area of 1.

57. Writing in standard form: $2x^2 + 4x - 3 = 0$

59. Writing in standard form:
$$(x-2)(x+3) = 5$$
$$x^2 + 3x - 2x - 6 = 5$$
$$x^2 + x - 11 = 0$$

61. The coefficients and constant term are: 1, –5, –6

63. The coefficients and constant term are: 2, 4, –3

65. Evaluating $b^2 - 4ac$: $b^2 - 4ac = (-5)^2 - 4(1)(-6) = 25 + 24 = 49$

67. Evaluating $b^2 - 4ac$: $b^2 - 4ac = (4)^2 - 4(2)(-3) = 16 + 24 = 40$

69. Simplifying: $\dfrac{5+\sqrt{49}}{2} = \dfrac{5+7}{2} = \dfrac{12}{2} = 6$

71. Simplifying: $\dfrac{-4-\sqrt{40}}{4} = \dfrac{-4-2\sqrt{10}}{4} = \dfrac{-2-\sqrt{10}}{2}$

8.6 The Quadratic Formula

1.   Using the quadratic formula with $a = 1$, $b = 5$, and $c = 6$:

$$x = \frac{-b \pm \sqrt{b^2 - 4ac}}{2a} = \frac{-5 \pm \sqrt{(5)^2 - 4(1)(6)}}{2(1)} = \frac{-5 \pm \sqrt{25 - 24}}{2} = \frac{-5 \pm 1}{2} = -3, -2$$

3.   Solving the equation:
$$a^2 - 4a + 1 = 0$$

$$a = \frac{4 \pm \sqrt{16 - 4}}{2} = \frac{4 \pm \sqrt{12}}{2} = \frac{4 \pm 2\sqrt{3}}{2} = 2 \pm \sqrt{3}$$

5.   Using the quadratic formula with $a = 1$, $b = 3$, and $c = 2$:

$$x = \frac{-b \pm \sqrt{b^2 - 4ac}}{2a} = \frac{-3 \pm \sqrt{(3)^2 - 4(1)(2)}}{2(1)} = \frac{-3 \pm \sqrt{9 - 8}}{2} = \frac{-3 \pm 1}{2} = -2, -1$$

7.   Using the quadratic formula with $a = 1$, $b = 6$, and $c = 9$:

$$x = \frac{-b \pm \sqrt{b^2 - 4ac}}{2a} = \frac{-6 \pm \sqrt{(6)^2 - 4(1)(9)}}{2(1)} = \frac{-6 \pm \sqrt{36 - 36}}{2} = \frac{-6 \pm 0}{2} = -3$$

9.   Using the quadratic formula with $a = 1$, $b = 6$, and $c = 7$:

$$x = \frac{-b \pm \sqrt{b^2 - 4ac}}{2a} = \frac{-6 \pm \sqrt{(6)^2 - 4(1)(7)}}{2(1)} = \frac{-6 \pm \sqrt{36 - 28}}{2} = \frac{-6 \pm 2\sqrt{2}}{2} = -3 \pm \sqrt{2}$$

11.   Using the quadratic formula with $a = 4$, $b = 8$, and $c = 1$:

$$x = \frac{-b \pm \sqrt{b^2 - 4ac}}{2a} = \frac{-8 \pm \sqrt{(8)^2 - 4(4)(1)}}{2(4)} = \frac{-8 \pm \sqrt{64 - 16}}{8} = \frac{-8 \pm 4\sqrt{3}}{8} = \frac{-2 \pm \sqrt{3}}{2}$$

13.   Using the quadratic formula with $a = 1$, $b = -2$, and $c = 1$:

$$x = \frac{-b \pm \sqrt{b^2 - 4ac}}{2a} = \frac{-(-2) \pm \sqrt{(-2)^2 - 4(1)(1)}}{2(1)} = \frac{2 \pm \sqrt{4 - 4}}{2} = \frac{2 \pm 0}{2} = 1$$

15.   Using the quadratic formula with $a = 1$, $b = -5$, and $c = -7$:

$$x = \frac{-b \pm \sqrt{b^2 - 4ac}}{2a} = \frac{-(-5) \pm \sqrt{(-5)^2 - 4(1)(-7)}}{2(1)} = \frac{5 \pm \sqrt{25 + 28}}{2} = \frac{5 \pm \sqrt{53}}{2}$$

17.   First write the equation as $6x^2 - x - 2 = 0$. Using $a = 6$, $b = -1$, and $c = -2$ in the quadratic formula:

$$x = \frac{-b \pm \sqrt{b^2 - 4ac}}{2a} = \frac{-(-1) \pm \sqrt{(-1)^2 - 4(6)(-2)}}{2(6)} = \frac{1 \pm \sqrt{1 + 48}}{12} = \frac{1 \pm \sqrt{49}}{12} = \frac{1 \pm 7}{12} = -\frac{1}{2}, \frac{2}{3}$$

19.   First simplify the equation:
$$(x - 2)(x + 1) = 3$$
$$x^2 - 2x + x - 2 = 3$$
$$x^2 - x - 5 = 0$$
Using $a = 1$, $b = -1$, and $c = -5$ in the quadratic formula:

$$x = \frac{-b \pm \sqrt{b^2 - 4ac}}{2a} = \frac{-(-1) \pm \sqrt{(-1)^2 - 4(1)(-5)}}{2(1)} = \frac{1 \pm \sqrt{1 + 20}}{2} = \frac{1 \pm \sqrt{21}}{2}$$

21. First write the equation as $2x^2 - 3x - 5 = 0$. Using $a = 2$, $b = -3$, and $c = -5$ in the quadratic formula:

$$x = \frac{-b \pm \sqrt{b^2 - 4ac}}{2a} = \frac{-(-3) \pm \sqrt{(-3)^2 - 4(2)(-5)}}{2(2)} = \frac{3 \pm \sqrt{9 + 40}}{4} = \frac{3 \pm \sqrt{49}}{4} = \frac{3 \pm 7}{4} = -1, \frac{5}{2}$$

23. First write the equation as $2x^2 + 6x - 7 = 0$. Using $a = 2$, $b = 6$, and $c = -7$ in the quadratic formula:

$$x = \frac{-b \pm \sqrt{b^2 - 4ac}}{2a}$$
$$= \frac{-6 \pm \sqrt{(6)^2 - 4(2)(-7)}}{2(2)}$$
$$= \frac{-6 \pm \sqrt{36 + 56}}{4}$$
$$= \frac{-6 \pm \sqrt{92}}{4}$$
$$= \frac{-6 \pm 2\sqrt{23}}{4}$$
$$= \frac{-3 \pm \sqrt{23}}{2}$$

25. First write the equation as $3x^2 + 4x - 2 = 0$. Using $a = 3$, $b = 4$, and $c = -2$ in the quadratic formula:

$$x = \frac{-b \pm \sqrt{b^2 - 4ac}}{2a}$$
$$= \frac{-4 \pm \sqrt{(4)^2 - 4(3)(-2)}}{2(3)}$$
$$= \frac{-4 \pm \sqrt{16 + 24}}{6}$$
$$= \frac{-4 \pm \sqrt{40}}{6}$$
$$= \frac{-4 \pm 2\sqrt{10}}{6}$$
$$= \frac{-2 \pm \sqrt{10}}{3}$$

27. First write the equation as $2x^2 - 2x - 5 = 0$. Using $a = 2$, $b = -2$, and $c = -5$ in the quadratic formula:

$$\begin{aligned} x &= \frac{-b \pm \sqrt{b^2 - 4ac}}{2a} \\ &= \frac{-(-2) \pm \sqrt{(-2)^2 - 4(2)(-5)}}{2(2)} \\ &= \frac{2 \pm \sqrt{4 + 40}}{4} \\ &= \frac{2 \pm \sqrt{44}}{4} \\ &= \frac{2 \pm 2\sqrt{11}}{4} \\ &= \frac{1 \pm \sqrt{11}}{2} \end{aligned}$$

29. a.   Solving by factoring:

$$3x^2 - 5x = 0$$
$$x(3x - 5) = 0$$
$$x = 0, \frac{5}{3}$$

b.   Using $a = 3$, $b = -5$, and $c = 0$ in the quadratic formula:

$$x = \frac{-b \pm \sqrt{b^2 - 4ac}}{2a} = \frac{-(-5) \pm \sqrt{(-5)^2 - 4(3)(1)}}{2(3)} = \frac{5 \pm \sqrt{25 - 0}}{6} = \frac{5 \pm \sqrt{25}}{6} = \frac{5 \pm 5}{6} = 0, \frac{5}{3}$$

31. Factoring out x results in the equation $x(2x^2 + 3x - 4) = 0$, so $x = 0$ is one solution. The other two solutions are found by using $a = 2$, $b = 3$, and $c = -4$ in the quadratic formula:

$$x = \frac{-b \pm \sqrt{b^2 - 4ac}}{2a} = \frac{-3 \pm \sqrt{(3)^2 - 4(2)(-4)}}{2(2)} = \frac{-3 \pm \sqrt{9 + 32}}{4} = \frac{-3 \pm \sqrt{41}}{4}$$

33. a and b are equivalent, since: $\dfrac{6 + 2\sqrt{3}}{4} = \dfrac{2(3 + \sqrt{3})}{4} = \dfrac{3 + \sqrt{3}}{2}$

35. Multiplying each side of the equation by 6 results in the equation $3x^2 - 3x - 1 = 0$. Using $a = 3$, $b = -3$, and $c = -1$ in the quadratic formula:

$$x = \frac{-b \pm \sqrt{b^2 - 4ac}}{2a} = \frac{-(-3) \pm \sqrt{(-3)^2 - 4(3)(-1)}}{2(3)} = \frac{3 \pm \sqrt{9 + 12}}{6} = \frac{3 \pm \sqrt{21}}{6}$$

37. Solving the equation:

$$56 = 8 + 64t - 16t^2$$
$$16t^2 - 64t + 48 = 0$$
$$t^2 - 4t + 3 = 0$$
$$(t - 1)(t - 3) = 0$$
$$t = 1, 3$$

The arrow is 56 feet above the ground after 1 second and after 3 seconds.

39. Adding the two equations:
$$5x = 10$$
$$x = 2$$
Substituting into the first equation:
$$2(2) + y = 3$$
$$4 + y = 3$$
$$y = -1$$
The solution is $(2, -1)$.

41. Multiplying the second equation by –4:
$$4x - 5y = 1$$
$$-4x + 8y = 8$$
Adding the two equations:
$$3y = 9$$
$$y = 3$$
Substituting into the second equation:
$$x - 2(3) = -2$$
$$x - 6 = -2$$
$$x = 4$$
The solution is $(4, 3)$.

43. Substituting into the first equation:
$$5x + 2(3x - 2) = 7$$
$$5x + 6x - 4 = 7$$
$$11x - 4 = 7$$
$$11x = 11$$
$$x = 1$$
Substituting into the second equation: $y = 3(1) - 2 = 3 - 2 = 1$. The solution is $(1, 1)$.

45. Substituting into the first equation:
$$2(2y + 1) - 3y = 4$$
$$4y + 2 - 3y = 4$$
$$y + 2 = 4$$
$$y = 2$$
Substituting into the second equation: $x = 2(2) + 1 = 4 + 1 = 5$. The solution is $(5, 2)$.

Chapter 8 Test

See www.mathtv.com for video solutions to all problems in this chapter test.